PESTS IN AND AROUND

the

FLORIDA HOME

by

Philip G. Koehler

I would like to thank the following individuals for their contribution to this book.

Production

Graphics and Design: Jane Medley
Desktop Publishing: Glinda Burnett
Editing: Susan Williams/Julia Graddy

Principal Contributors

R. H. Baker -- *Human Malaria*
J. L. Castner -- *Photography on Color Sheets*
N. C. Hinkle -- *Reduced Chemical Control of Fleas*
W. H. Kern, Jr. -- *Reduced Chemical Control of Fleas, Rats and Mice, Non-Chemical Rodent Control, Bats and Birds*
C. A. Morris -- *Human Malaria and Eastern Encephalitis - A Fatal Mistake*
J. K. Nayar -- *Human Malaria*
F. M. Oi -- *The Formosan Subterranean Termite*
M. T. Sanford -- *Removing Honey Bee Nests and Information On The African Honey Bee*
R. H. Schreffrahn -- *The Formosan Subterranean Termite*
D. E. Short -- *Stinging and Venomous Insects, Color Sheet Venomous Spiders in Florida and Color Sheet Stinging and Venomous Arthropods*
D. R. Suiter -- *Cockroach Control Without Insecticides*
N. Y. Su -- *The Formosan Subterranean Termite*

For more information about books and other publications available from The University of Florida, contact IFAS Publications, 904-392-1764, P. O. Box 110011, University of Florida, Gainesville, FL 32611-0011.

For information about specific controls, consult the Florida Insect Control Guide, SP134, or contact your County Cooperative Extension Service office.

TABLE OF CONTENTS

Introduction

Urban Pests and Pest Management

URBAN PESTS

Florida is an ideal environment for a wide variety of pests. Some of these pests live and reproduce inside structures; however many live outdoors and only occasionally invade the home or workplace. Because almost everyone has problems with pests, most urban areas are sprayed with pesticides. Most pesticide applications are unnecessary and can result in environmental contamination and human exposure to pesticides. Approximately 30 to 40% of the pesticide use in Florida is in urban areas. Considering the high concentration of population and the sensitive environment, safety and risk of exposure to pesticides is disproportionate in our fragile urban ecosystem. Do not apply control measure until a pest population is present and damage is beginning to occur. Judicious use of pesticides and the implementation of integrated pest management practices is of utmost importance.

PEST MANAGEMENT

Pest management is a decision-making process that involves locating and monitoring pests, establishing thresholds for action and selecting pest control methods. To do this, the habits and life cycles of many pests must be understood and appropriate measures to solve pest problems must be implemented.

One important integrated pest management practice commonly used in the urban environment is prevention of pest problems. Managing pests through prevention is usually less expensive than trying to control a pest population that has already become established. Furthermore, pest prevention reduces the chance for substantial economic loss or irreversible damage. Prevention avoids some of the disruption associated with control efforts that may be needed after pests become established.

Once a pest becomes established, the most common pest management goal is to eliminate it. Elimination can only be successful if the conditions that originally favored the pest can be modified or pest entry can be completely blocked.

Locating and Monitoring Pests

Decisions to use pesticides and other control methods should be based in part on pest detection and monitoring results. Visually inspecting an area where pests or their damage is observed is the most common method of detection. Inspection involves careful and thorough searching for signs of the pest and conditions that favor pest buildup. Monitoring is a systematic method of observing pests or pest signs over a period of time. Monitoring may help you detect unwanted pests and determine where pests are coming from and where they are living. Monitoring is also helpful in evaluating control programs. Special devices and tools are available to detect and monitor certain types of pests.

Visual Inspection

The purpose of a visual inspection is to search for evidence of pests. During an inspection, look for: (1) conditions that favor pests; (2) signs of pest damage, entry, or presence; and (3) the pest itself.

When doing an inspection, it's helpful to prepare sketches of the structure, lawn or garden. Observe any conditions that may cause problems during pest control operations. Note areas that you were unable to inspect because they were inaccessible. Show the locations of trees, shrubs, trash and garbage storage, water sources, and other features of the surrounding area that may attract or harbor pests or promote pest buildup.

Detection and Monitoring Devices

Different types of simple devices can assist you in detecting and monitoring many of the pests found in structures. These include pheromones and other attractants, light traps, flypaper, and sticky traps.

Pheromones and Other Attractants

Pheromones are chemicals normally produced by certain insects (and other animals) to affect the behavior of individuals of the same species. Pheromones are used by insects for mating, aggregation, feeding, trail following, and recruitment. Synthetically made pheromones mimic the action of pheromones produced by some pest insects. These are useful for monitoring the adult forms of pest moths, certain beetles and weevils, and some species of flies and fruit flies. Certain other materials are also used as trap attractants. For example, ammonium carbonate attracts many different species of flies; food-like odors attract certain insects.

For monitoring, pheromones and other attractants are used in sticky traps or rodent spring traps. Inside a building where food is stored, you can use these attractant traps to locate sources of infestation. The effectiveness of attractant traps is influenced by the number of traps used and where they are placed.

Check traps regularly. For insects, check traps once or twice per week at a minimum and remove all captured insects. Clean or replace sticky surfaces whenever they become covered with debris.

Record the number of target insects removed from traps each time they are checked. Plot trap catches on a per-day basis. This will allow you to perceive changes in the insects activity and verify the success of control measures. Compare this activity with activity in traps in other locations.

Light Traps

Traps equipped with ultraviolet lights, or black lights, attract several species of flying insects. These traps usually have a container with a funnel-shaped entrance that allows insects to enter easily but blocks their escape. Some light traps have an electrically charged grid that kills insects as they approach the light. Electrocuter traps are usually not used for insect monitoring.

Flypaper

You can use flypaper for monitoring flying insects within confined areas. Some manufacturers add a fly pheromone to the sticky coating to make it more effective.

Examine the captured insects to determine their identity. Keep records of the numbers and species of pests that were caught and use this information for selecting and evaluating control methods.

Sticky Traps

For monitoring cockroaches, place glue boards along travelways next to intersections of walls and floors. Place them in cabinets and next to major appliances in the kitchen.

Establishing Thresholds For Action

Pest control decisions are influenced by health or safety dangers created by the pest, by legal restrictions on pest infestation, and by levels of pest tolerance. Occasionally a pest control decision depends on the costs involved to control a pest weighed against the benefits received. On the basis of any of these factors, a threshold for action can usually be established to determine what type of control is needed and when control should begin.

Health and Safety Threshold

Health or safety threats commonly require fast, extensive, and sometimes costly pest control measures. Several pests have the potential for causing injuries to people (mosquitoes, biting bugs, fleas, spiders, bees, and wasps, for example) or transmitting diseases to people or animals (rats and mice, cockroaches, fleas, flies, and mosquitoes). Some others, such as termites, and wood-boring beetles, cause the type of damage that makes structures unsafe or reduces their value.

Decisions to control pests are based on knowledge of the potential harm they can cause. If serious injury or damage may result, the control threshold must be very low. For instance, one rat chewing on electrical wiring can cause a serious fire.

Legal Thresholds

Public safety codes often require control of pests in public buildings, commercial housing, food service facilities, and other public structures. Building and safety standards address the control of structural pests as well as the repair of damage caused by them. These legal thresholds dictate when pest control methods must be used, even though in some cases control methods cannot be economically justified or the pests may not be causing a hazard to public health or safety.

For information on laws that regulate pest infestation in certain buildings and on foods, contact state and local health departments and housing and community development offices. The Bureau of Entomology has information on laws that apply to the control of structural pests. Federal marketing orders list allowable tolerances of specific pests or pest damage in fresh and stored food items; this information can be obtained from the United States Department of Agriculture, Agricultural Marketing Service.

Pest Acceptance Threshold

People have different degrees of acceptance of pests that they are willing to tolerate in and around their homes. Pest acceptance thresholds may be high because of social or cultural factors or because of concerns about the costs or hazards of pest control methods used. A pest acceptance threshold can be extremely low due to a person's revulsion or fear of the pest. Acceptance thresholds may sometimes be modified if you can provide factual information about specific pests, the potential for pest damage, and methods of pest control.

Economic Threshold

In certain instances, the cost of control measures may need to be justified. Economic thresholds may apply if there are no health and safety, legal, or tolerance thresholds that need to be considered. An economic threshold is a level of pest abundance at which the potential loss caused by pest damage is expected to be greater than the cost of controlling the pest.

INTEGRATED PEST MANAGEMENT METHODS IN HOUSEHOLDS AND STRUCTURES

Pests can be prevented, through sanitation and habitat modification, or they can be controlled by trapping, pesticide use, and, in some instances, biological control. Pests in structures are usually more effectively controlled when a combination of compatible control methods can be used.

Sanitation and Habitat Modification

Habitats are areas within the larger environment that are suitable for a pest's survival. Habitats provide a pest with some or all of its necessary living requirements. A habitat can only accommodate a maximum number of pests due to limitations of one or more of these requirements. This maximum number is known as the carrying capacity. Where large quantities of food are available and shelter and other requirements are ample, the carrying capacity is high. Such a habitat can support an almost unlimited number of individuals of a pest species. If the carrying capacity is limited, however, the population tends to remain fixed in size. If you remove individuals from a habitat through pest control measures or if they die off due to natural causes, these individuals will be replaced by others, usually soon, unless the carrying capacity is reduced at the same time. Population size is maintained at the carrying capacity by increased reproduction among remaining individuals or by new individuals migrating in.

Habitat modification usually involves improving sanitation practices. Sanitation includes removing food, water, breeding sites, and shelter used by pests. Outdoors, you may need to trim or remove dense, pest-harboring vegetation near buildings, clean up trash, keep garbage in closed containers, provide for drainage of standing water, clean up animal wastes and spilled animal feed, and eliminate items that attract pests. Inside, sanitation includes storing foods and food wastes in tightly closed containers, cleaning up spills and residues, removing trash and other materials that can be used for nests, and thorough vacuuming and dusting on a regular basis. The cleaning of surfaces may also improve the effectiveness of pesticides by removing grease, oils, dust, and other contaminants that interfere with their function. To assist in good sanitation, make sure interior areas are well lighted to simplify cleaning and easy detection of pests and pest damage. Sweepings and other wastes should be taken to a disposal area outside of the building.

Other sanitation practices include removing dirt mounds, wood pieces, and other cellulose debris from areas beneath buildings to keep from promoting termite problems. Provide adequate ventilation to areas beneath buildings to reduce moisture.

Outdoor lights placed near entrances to buildings attract many flying and crawling insects at night. If possible, locate light fixtures away from entrances. Otherwise, modify the type of light being used. Sodium vapor lights are better than mercury vapor lights or standard incandescent lights for outdoor use because they emit a spectrum of light that is less attractive to insects; yellow "bug" bulbs work on the same principle.

A program of sanitation and habitat modification requires cooperation. All people living or working in a building must keep food, food waste, and trash in pestproof containers and store other items in designated places where they cannot attract pests. Inhabitants should promptly report pest problems. Housekeeping and landscape maintenance workers can help by keeping interior and exterior areas free of trash, nesting sites, and other items that might be attractive to pests; they should provide containers for wastes and specify locations for storage of other materials. Buildings must be monitored on a regular basis to ensure that sanitation conditions are maintained and to spot new problem areas as they occur. Tenants and persons responsible for housekeeping and landscape maintenance must be notified of conditions that promote pest buildup so they can take corrective action.

Exclusion

Exclusion is a type of habitat modification useful for keeping fleas, ants, cockroaches, stored-product pests, termites, and other pests from entering buildings. The design and construction of a building may either promote pests or exclude them. Pestproof design and construction, should be an important consideration when planning new structures and remodeling older ones.

Check building exteriors for ways that insects, or other pests can enter. Obvious entrances for many types of pests are doorways and windows. These must be fitted with tight-fitting screens or doors. Properly installed weatherstripping eliminates small cracks that provide access for some pests. Also, check attic and foundation vents to ensure that they are tight and screened. Look for foundation or wall cracks, gaps in

siding or joints, and areas where pipes, wires, or other objects pass through walls. Fill openings with concrete or another suitable patching material. Inspect chimneys and roof vent pipes for adequate screening.

Inspection

Inspect items brought into a building for pest infestation. For example, firewood may harbor carpenter ants, spiders, cockroaches, wood-boring beetles, termites, or similar pests, or eggs of some pests. Furniture, rugs, and other items moved from an infested building can be contaminated with cockroaches, carpet beetles, or fleas. Dogs and cats bring in fleas and ticks.

Trapping

Besides their benefits as monitoring devices, traps are used to kill pests or to catch pests so they can be removed from an area. Many types of pests can be controlled through trapping. Traps do not require the use of potentially hazardous chemicals, and the user can easily view the success of the trapping program. However, successful trapping programs require skill, time, and attention to develop workable techniques. Even so, trapping may not always work well enough under some conditions to satisfactorily control target pets. Trapping techniques that are successful in one situation may not always work as well under different conditions or at other locations.

Traps include sticky traps, pheromone traps, and light traps.

Biological Control

Biological control is gaining more importance as a pest control method for certain insects in structures. Cockroach populations have been successfully reduced in certain locations by introducing parasitic wasps. Biological control techniques either augment other control practices or replace more disruptive or hazardous methods.

Pesticides

The application of pesticides is the most common pest control method used in and around buildings, enclosed areas, and vehicles. Some pesticides provide chemical barriers to prevent insects from getting in.

6

Pesticides are also used to treat soil, wood, fabrics, and other items to prevent pest damage.

Pesticides are available as baits, tracking powders, desiccants (inert dusts or sorptive powders), liquids, dusts, and gases. The type of pesticide used and the kind of formulation selected is based on the life habits of the pest, its density, and its location.

SUMMARY

This manual on pests in and around the home, is a compilation of information on biology, identification, and pest management practices for urban pests. By learning about these specific pests it should be possible to reduce pesticide use in the home. The reduced pesticide use should improve the health and safety of our families and the Florida environment.

Using Pesticides Safely

When applying any pesticide, you assume the legal responsibility for using it strictly in accordance with label instructions. You must always protect people who live or work in the treated area so they are not exposed to harmful residues. Avoid using pesticides or application methods that might injure nontarget animals or plants or damage property. Pesticide use should not endanger the environment or cause contamination of groundwater, soils, air, or human and animal foods. In addition, you must use pesticides in ways that avoid excessive exposure to any part of your own body. Precautions that must be observed when handling pesticide containers, including guidelines for mixing pesticides, as well as, some of the steps that must be taken to properly apply pesticides, ways to safely store these materials, and information on pesticide disposal are summarized.

LIQUIDS

Pesticide liquids are mixtures of powdered or liquid active ingredients combined with liquid carriers such as water or petroleum products. Pesticides may dissolve in the carrier to form a solution or may remain suspended in the liquid to form an emulsion or suspension. Suspensions and emulsions require constant agitation to maintain a uniform spray mixture.

Liquid pesticides are applied as spot treatments, crack and crevice treatments, fogs or mists in confined areas, or general sprays to large areas. The common ways to apply liquid sprays are with aerosol dispensers, hand-held compressed air sprayers, backpack sprayers, or larger, motorized spray units.

When liquid sprays are applied, a residue of pesticide active ingredient remains on the treated surfaces and helps to control pests over a period of time. The length of time depends on the type of pesticide used, the type of formulation, the concentration of active ingredient applied, the type of surface treated, and environmental influences such as temperature, humidity, or sunlight.

Undiluted pesticides contain concentrated amounts of active ingredient that may cause serious injury if inhaled, splashed or blown into the eyes, or spilled on the skin or clothing. Some concentrated pesticides may be flammable.

Applying liquid sprays in certain areas may be extremely hazardous. For example, electric outlets, motors, or exposed wiring pose a potential threat of electrical shock to persons applying water-based pesticide sprays. Pilot lights and gas flames from heaters and appliances may ignite flammable petroleum based pesticides. Sparks from electric motors and switches and glowing heating elements may also ignite flammable materials. Pesticide vapors or fumes in confined areas may injure people if ventilation is inadequate.

GASES

Gases that kill pests are known as fumigants. Fumigants are used to control certain stored-product insects, drywood termites, wood-destroying beetles, soil-infesting nematodes, soil pathogens, and some rodents. The process of applying fumigants, or fumigation, is much different from other forms of pesticide application and requires application by a professional pest control operator.

DUSTS

Dust formulations are finely ground dry powders that contain toxic materials. These are sometimes used to control certain insects. Most dusts are blown into inaccessible places where pests hide. Dusts do not penetrate surfaces, and they usually break down slowly. Therefore, the active ingredient in dust formulations remains on the treated surface and is active against pests for a long period of time if the treated area stays dry. Because they do not penetrate, dusts are more effective than liquids on absorptive surfaces such as concrete.

Dusts may be applied in cracks and crevices, under cabinets or appliances, and in other areas inaccessible to children and pets. This formulation leaves visible residues on treated surfaces, which often limits its use to areas such as warehouses, attics, crawl spaces, and wall voids.

Dusts usually provide better coverage than sprays in inaccessible or hard-to-reach places. In wall voids, they can be dispersed with compressed air to better reach all surfaces. During manufacture, dusts are sometimes given an electrical charge or they are combined with an electrically charged powder to make them cling to surfaces better. Bulb applicators, shaker cans, aerosol cans, and compressed air dusters are used to apply dust formulations.

When using dusts, prevent their drift into the airspace of rooms or work areas. Apply dusts only according to the instructions on the pesticide label. Wear approved respiratory protection to avoid inhaling dust particles.

Desiccants are dusts or sorptive powders used to control certain insect pests found in buildings. The powder abrades or adsorbs the waxy coating that protects insects from losing body water. Desiccants often last longer than other forms of insecticides; however, insects must move through the material or dust and pick some up on their bodies for it to be effective. Desiccants should be blown into wall voids, attics, and crawl spaces and also into other areas where insects hide. Some desiccants are repelling, which helps keep insects from treated areas. Avoid breathing dusts during application by wearing respiratory protection.

GRANULES

Granular formulations are sometimes used to control ants, sowbugs, earwigs, snails, slugs, and occasionally other soil-inhabiting organisms. Usually granules are combined with a food substance or attractant to encourage target pests to feed on them. Do not apply granules in areas where children or pets may come in contact with them.

Poisoned Bait

Poisoned bait may be used to control specific types of insects. Most baits are a combination of a pesticide and a food material. Baits are usually placed in a bait station or broadcast over the soil around the outside of a structure.

Choose bait types and bait station styles on the basis of (1) the type of pest being controlled, (2) the past history of bait use, and (3) the conditions where baiting will take place. For example, when baiting for ants, select a bait that foraging workers will carry back to the nest to feed to the colony's reproductives and brood; the toxic substance must be slow-acting so that foraging workers are not killed before they reach the nest. Bait used to control flies, on the other hand, must be fast-acting in order to stop continued annoyance and prevent further egg laying.

Place or apply insect baits in areas of greatest activity or in areas that cannot be sprayed or dusted. To treat ants, place the bait; along trails, near nest entrances, around the foundation of the building, and under sinks and other out-of-the-way locations inside the building. Apply cockroach bait under appliances, under sinks, behind furniture, and in hidden areas where these insects have been observed or are suspected to occur. Place bait at wall intersections as cockroaches tend to travel along edges. For cockroach species that occur outdoors, place baits in or around woodpiles, in water meter boxes and other protected locations where these insects are usually found.

HOW PESTICIDES CAN INJURE PEOPLE

Poisonous chemicals such as pesticides injure or kill people by interfering with the normal functioning of internal body organs and systems. The nature and extent of injury depends on the toxicity of the chemical as well as the dose (amount of material) that enters the body. A person's health and size may also influence the severity of injury.

The ingredients of some pesticides are very potent and are capable of causing poisoning at doses as small as a few drops. Other less-toxic pesticides might require as much as several pounds be consumed before signs of illness appear. Regardless of the specific potential hazard, anyone working with pesticides should avoid exposure by using suitable protective clothing and application techniques. Anyone living or working in pesticide-treated areas must also be protected from exposure levels that will result in injury.

Poisoning Symptoms

Symptoms are abnormal conditions, feelings, or signs that indicate the presence of an injury, disease, or disorder. When a person is exposed to a large enough dose of pesticide to cause injury or poisoning, some type of symptoms will usually appear. These symptoms may show up immediately or after several days; sometimes they may not appear until after several months or years. It may be difficult to associate the illness or injury with its cause if there has been a lapse of time between exposure and observable effect.

The effect of an exposure can be localized, such as eye or skin irritation, or generalized, when the pesticide is absorbed into the blood and distributed to other parts of the body. A pesticide can affect several different internal systems at the same time. If the person experiences an injury but recovers quickly, or gets worse and dies within a short time, it is known as an acute illness or injury. If the resulting effects last for a long time, and perhaps are irreversible, it is known as chronic. Examples of chronic conditions usually associated with high or prolonged levels of exposure to certain pesticides include, among others, infertility, birth defects, and cancer. Pesticides that are found to cause such disorders or are suspected of causing these problems may lose their federal registration and can then no longer be used in the United States.

Some pesticide poisoning symptoms are similar to symptoms produced by many other chemicals or conditions. The type of symptoms may vary between chemical classes of pesticides and may also be different among pesticides within the same chemical class. The presence and severity of symptoms usually are proportional to the amount of pesticide (the dosage) entering the tissues of the exposed person. Symptoms may include a skin rash, headache, or irritation of the eyes, nose, or throat. These symptoms disappear within a short period of time and sometimes are difficult to distinguish from symptoms associated with an allergy, cold, or the flu. Other symptoms, which might be caused by higher levels of pesticide exposure, include any of the following: blurred vision, dizziness, heavy sweating, weakness, nausea, stomach pain, vomiting, diarrhea, extreme thirst, and blistered skin. Poisoning can also result in apprehension, restlessness, anxiety, unusual behavior, shaking, convulsions, or unconsciousness of the victim. Although these symptoms can indicate pesticide poisoning, they also may be signs of other physical disorders or diseases. Whenever the possibility of poisoning exists, consult a physician; be sure to have readily available a copy of the pesticide label or the name of the pesticide, the manufacturer, and the EPA registration number. Diagnosis of a pesticide related injury usually requires; careful medical examinations, laboratory tests, observation, and familiarity with a person's medical history.

Individuals commonly vary in their sensitivity to pesticides. Some people show no reaction to a dose that can cause severe illness in others. A person's age and body size may influence their response to a given dose, thus infants and young children are normally affected by smaller doses than adults. Also, adult women may be affected by smaller doses of some pesticides than adult men. The unborn child carried by a pregnant woman may be highly sensitive to exposure to some pesticides by the perspective mother.

Pesticides that are applied in strict accordance with their label instructions and with adherence to application rates, reentry intervals, protective equipment requirements, aeration periods, and other listed procedures- generally do not leave unsafe levels of pesticide residues. Accidents during application may result in a higher, and sometimes unsafe, exposure. An improper application caused by not following label instructions may also result in injury.

PROTECTING PEOPLE

Always apply pesticides in strict accordance with label instructions. Never use a pesticide in a building or other area unless people living or working there can be protected from exposure. This often requires that inhabitants leave the area before an application begins and that they remain away for a specified period of time after the application has been completed. To reduce personal exposure, remove or cover food and utensils before pesticide applications are made. Protect linens, bedding and similar items, open windows and doors in order to increase ventilation after an application has been made. Vacuum carpets and clean floors after an treatment, and keep children and pets away from these areas.

Insecticides may be needed to control pests in places where food is stored, prepared, or eaten. If so, special precautions must be taken. For instance, never treat food preparation surfaces with dusts or liquid sprays and do not allow residues to drift onto food or utensils. If fogs are used, all food preparation surfaces must be thoroughly cleaned after application.

Never make an application near air ducts or ventilation systems unless the system can be shut down for a period of time. Do not apply pesticides inside heating or cooling ducts.

Infants, Children, the Elderly, and People with Medical Conditions

Sometimes the use of pesticides must be restricted or avoided to protect people living in the targeted area. Rely on nonchemical control methods as much as possible, and use a pesticide only where absolutely necessary. When pesticides are needed, choose the safest formulation available such as a bait or a liquid spray having low volatility and follow all label instructions and precautions. Be extremely careful when using pesticides in areas occupied by infants, children, the elderly, or a person who is sick. These areas include hospitals, nursing homes, schools, and certain households.

Infants are more vulnerable to pesticide exposure than larger children or adults. This is because of their small size and undeveloped immune system that are responsible for detoxifying hazardous chemicals. Do not apply a pesticide to any item used for infant care, and avoid spraying or dusting carpets, clothing, blankets, towels, or any fabrics that infants or others may contact. When a pesticide is needed in areas where an infant may spend part of the day, use a formulation that will break down completely before the infant returns.

Children under the age of six are active and curious and it is difficult to keep them away from places where a pesticide has been used for control of household pests. Young children are highly mobile and active exploring and put many objects (including their hands) into their mouths. They also frequently crawl on floors and climb on other surfaces. Therefore, never apply a pesticide to play equipment, toys, or any surfaces normally contacted by the youngsters. On carpets, use pesticides that break

down rapidly. In all cases, use pesticides having low toxicity and low volatility. If you use bait stations or traps, secure them in a place well out of reach and out of sight.

Elderly people may be susceptible to respiratory illnesses and other physical disorders that may result in them having them a low tolerance to many airborne dusts and chemicals, including certain pesticides. In some instances, their bodies may be unable to properly degrade or eliminate foreign or toxic materials, such as pesticides. Therefore, use extreme caution when making pesticide applications in rooms where elderly people sleep or spend long periods of time and whenever possible, avoid treating these places. In other areas, use a pesticide with low toxicity and low volatility and spot treat as much as possible to reduce potential hazards. Select alternate methods of control whenever possible, and always augment pesticide use with other pest control techniques so that the amount of pesticide used can be minimized.

People who are acutely ill or suffer from conditions such as diabetes, alcoholism or have allergies or respiratory disorders including asthma and emphysema may be more sensitive to pesticides in their environment. Medications used to treat illnesses may influence the effects of pesticide exposure. Provide persons who are ill or using medications with the name of the pesticide you plan to use and ask them to contact their physician for advice.

APPLICATOR SAFETY

Safety risks for applicators working in buildings or enclosed areas are compounded by hazards such as electrical equipment, possibility of explosions, and confined work areas. Learn to recognize hazards in the application site that could cause injury. Avoid pesticide exposure by wearing required or recommended protective equipment. Carefully maintain, clean, and store protective equipment in order to keep it in good condition and to ensure that it provides optimum protection.

Fire, Explosion, and Electrical Hazards

Fires, explosions, and electrical hazards can occur in residential, industrial, and institutional settings and other confined areas. Before using a pesticide, examine the intended application site for hazards. For example, never apply a pesticide dissolved in oil or petroleum solvent in an enclosed area if there is any source of spark or flame such as functioning electrical motors, wall switches, appliances, or pilot lights. Before making an application, shut off electric and gas services to the treatment area. Avoid the use of aerosols in wall voids near hot water pipes since heat from these pipes can ignite solvents and cause a fire. Do not use dust in an enclosed area if there is an ignition source. Any airborne dust at the right concentration can explode. Boric acid dust is capable of extinguishing a pilot light, which could create an explosion hazard due to escaping gas (most new gas appliances are equipped with safety shut-off devices or igniters in place of pilot lights).

Do not use a water-based spray around electric appliances, outlets, or switches unless the power has been shut off. Water conducts electricity, so you are at risk of electrocution if the spray touches a live power source.

Working in Confined Areas

Confined areas present special hazards to persons making a pesticide application. Confined areas may be attics, crawl spaces beneath buildings, storage areas, closets, small rooms, and other places that have poor ventilation. Hazards include inhaling the pesticide being applied and coming in contact with treated surfaces. Cramped areas also may be uncomfortably hot due to poor air circulation. High temperatures may increase the applicators exposure potential, because sweating accelerates the rate of skin absorption of some pesticides.

Exposure hazards should be reduced when working in confined areas by wearing personal safety equipment. Whenever possible, increase ventilation in the treatment area by opening windows or using a fan to bring in fresh air. Always begin the application from a point furthest from the exit and never walk or crawl through freshly applied pesticide.

To avoid breathing fumes, wear an approved respirator for the pesticides being applied. Be sure it is in good working condition, fits well and thoroughly forms a good seal well around your face.

Prevent skin or eye contact with spray residues or vapor. When making an application, always wear a longsleeved shirt and full-length pants, coveralls, or lightweight spray suit. Protect your hands with waterproof gloves and use a faceshield or goggles to prevent spray or dust from getting into your eyes. Read the pesticide label carefully for the minimum protective clothing requirements.

Protecting Pets and Domestic Animals

Pets housed in or near residences or other buildings that are to be treated can often include several types of mammals, birds, reptiles, amphibians, and fish. Associated with pets and domestic animals are their food and water supplies, bedding, pens, equipment, and toys.

Most animals are susceptible to injury by pesticides. Some types of pesticides that are applied at low doses. Fish and birds are among the most susceptible. Cats are very sensitive because they are metabolically unable to detoxify many types of pesticides. Young animals as well as older or sick animals may be affected by lower pesticide doses than adult or healthy animals. Cats and dogs often lie and sleep on the ground and other surfaces that may have been treated and then they may groom and clean themselves by licking. This process can increase their potential for exposure even when small amounts of pesticide have been used.

In order to provide protection for pets and domestic animals, remove them from the area before making a pesticide application. Keep animals away until the spray dries and the area is well ventilated. Do not apply pesticides on or near animal food or water or dishes in which are used in feeding. If the animals are returned to the treated area, flea collars should be removed and any ectoparasite systemic medications should be discontinued.

Protect pests in aquariums. First turn off aerators or unplug them so airborne pesticide is not bubbled into the water. Second, cover the tank with newspaper or plastic bags to prevent sprayed pesticide from drifting into the water. Third, place any food or drugs used to treat the pets in plastic bags so they do not become contaminated with insecticide. Protect birds or other animals in cages by asking the owner to remove the pet from the premises or carrying the cage to an untreated bedroom. Keep the animal out of sprayed rooms until the treatment is dry.

Pesticide Drift

If pesticides are not carefully applied, they may drift beyond the treatment site and become deposited as unacceptable residues on surfaces not intended to be treated. These residues can possibly endanger nontarget organisms. Residues from improper application or improper rinsing of equipment may also result in contamination of surface or groundwater.

Preventing Drift or Unwanted Exposure

Do not use dust formulations in outdoor locations since they easily drift to areas not to be treated. To prevent drift when applying liquid sprays, use low pressures and large nozzle orifices. This reduces formation of small droplets that are subject to drift. Never make an outdoor application of a liquid spray when the wind is blowing faster than 5 miles per hour. If there is a slight wind, select a formulation or adjuvant that reduces drift. Be especially careful if you are spraying near fruit trees or vegetable gardens, flowers, clothing being air dried outside, cars, windows, and dark surfaces that may spot. Special care should be practiced around pet or livestock food and water containers, fish ponds, bird baths, swimming pools, saunas, spas, or outdoor furniture. Avoid outdoor applications that may drift to children's play areas, sandboxes, swing sets, or lawns and shrubbery that children contact.

Do not apply a pesticide in outdoor locations where residues can be carried into a well, stream, pond, or other water source. Never drain or wash application equipment where runoff will enter sewers, sinks, sumps, or drain tile systems.

When making a liquid or dust application inside a structure, keep the spray or dust away from air ducts, fans, or blowers in order to prevent the material from being blown into non-target areas.

CHARACTERISTICS OF TREATED SURFACES

Treatment sites may have surfaces whose characteristics must be evaluated before applying a pesticide. Depending on the type of surface, a pesticide can be absorbed and rendered ineffective, or the surface may be stained or etched. Concrete, for example, is porous and tends to absorb liquid sprays, reducing the amount of residue on the surface that is available to control target pests.

Floor coverings such as linoleum, tile, and carpeting can be stained or etched by some pesticides or solvents. Certain wallpapers and carpets contain dyes that may run, dissolve, fade or change colors if exposed to components of some pesticides. Paint and other finishes used on walls or woodwork may also react with spray chemicals to produce spotting or discoloration. Fabrics of all types, and the dyes used to make their patterns and color, may also react, affecting future wear or causing a stain or change in color. A soiled fabric may react differently than a clean one. Fabrics also can absorb a liquid pesticide, reducing pest control effectiveness.

Dust formulations can leave an unsightly residue if applied to surfaces of furniture, woodwork, fabrics, and other items in the treatment area.

Preventing Problems

Stains or color changes may be caused by an excessive dose or by certain application techniques. The formulation type being used may affect staining or spotting. A soiled or greasy surface may increase staining, spotting, or absorption. Paint that has been recently applied and not fully dried or cured has more of a tendency to spot.

Whenever possible, first apply a pesticide to an inconspicuous area, such as a closet, and allow the pesticide to dry for several hours to observe the reaction. Care should be taken when treating upholstery, furniture, drapes, or lower wall surfaces with a pesticide (lower wall surfaces are more likely to be soiled, which may enhance staining or bind the pesticide to make it less effective). Read and follow label directions and precautions carefully to avoid: staining, spotting, visible residues, and pesticide deactivation. Thoroughly clean the application equipment before adding a pesticide to prevent a possible reaction between the pesticide and leftover contaminants in the equipment. These contaminants may cause stains or other adverse effects.

When two or more pesticides are mixed, additional problems associated with pesticide compatibility may appear. Check the compatibility of pesticide mixtures before application by mixing a small quantity to determine whether separation or discoloration occurs.

Odor Problems

Many pesticides have odors that can be detected during and after application. Odors are usually strongest when pesticides are first applied. In confined areas, odors may become overpowering and objectionable; they can cause nausea or headache, initiate asthma or other breathing difficulties, or may trigger other medical or anxiety-related symptoms.

An odor may be a chemical characteristic of the pesticide its solvent, or it may be a substance added to the pesticide as a warning agent to reduce chances of injury. Reduce problems associated with odors by (1) using only the application rate stated on the pesticide label, (2) applying the pesticide in localized areas or as a spot treatment whenever possible, (3) using a low-odor formulation if available and if appropriate, (4) increasing ventilation to the application area by opening windows and doors or using fans, and (5) applying the pesticide during periods when the building is not occupied.

An odor may also be caused from a reaction between the pesticide and surfaces that have been treated. Before applying any pesticide in a confined area, read the pesticide label to determine if any of the chemicals in the formulation will react with treated surfaces to produce an odor.

How To Buy Pest Control Services

Pest control, like any kind of business, is a joint venture between the company and the customer. As with many businesses, the customer may have difficulty judging whether he has received the most skilled efforts for his money. This publication should help customers objectively look at the, service they are buying to determine its merit.

ORGANIZATION OF PEST CONTROL COMPANIES

Pest control companies are operated by people who usually have been assigned to specific responsibilities within a firm. The general categories for personnel in a pest control firm are: business license holder, manager, certified operator, salesmen, and service technicians.

The business license holder is the person who is licensed by the Florida Department of Agriculture and Consumer Services, Bureau of Entomology and Pest Control, to operate a pest control business. The business license holder is responsible for records, reports, advertising, and personal protective measures. All records of pest control activities, including contracts, must be kept at the business location or at a specified location registered with the Department of Agriculture and Consumer Services (DACS). The license holder also takes responsibility for the accuracy of advertising and for protecting employees with necessary equipment for safe pesticide application. Licenses for pest control companies are only available from DACS. Without the Department's license it is illegal for pest control service to be performed as a business.

The certified operator is in charge of the pest control activities of the business license holder. The certified operator is the real professional in the company who has demonstrated competence in pest control and supervision of such work. Certified operators must pass a rigid examination to demonstrate competence. In order to qualify to take the examination, they must have a minimum of 3 years of documented employment in pest control or a college degree with advanced training or a major in entomology, ornamental horticulture, or pest control technology. Certified operators may become certified in 4 categories: general household pest control including rodent control, termites or other wood-destroying organisms, lawn and ornamental pest control, and fumigation. Companies which have operators certified in all 4 categories may offer total pest control services; otherwise they may offer only the services for which they have a certified operator.

In the company, the certified operator is responsible for the training and supervision of personnel in the use of methods and equipment which are known to provide the best pest control. He selects the pesticides and assigns only properly trained and qualified personnel to perform pest control work. All safety practices such as safe storage, application, and disposal of pesticides are his responsibility. The quality of the service technicians in the company are a reflection of the quality of the certified operator.

The salesman is hired to inspect premises, determine the extent of the problem, and give estimates for pest control work to be performed; the salesman is sent to "sell" the company and its services to the customer. This is often a free service and customers should "shop around" to be certain the diagnosis of the problem is correct and the estimate is reasonable.

The service technician is the individual who does the actual pest control work. The quality of the control of pests depends to a large degree on the training the technician has received and the professionalism of the technician on the job.

All pest control employees must carry a company identification card issued by DHRS and must show it to you upon request.

HOW TO SELECT PEST CONTROL SERVICES WISELY

Your first contact with a pest control company may be through a salesman. A salesman is sent to sell you pest control services. Salesmen are ambitious and have been known to tell customers that a specific service is necessary when none is required. Selling unnecessary pest control services is prohibited by the pest control laws. However, homeowners must buy pest control services wisely.

The following steps should be taken by homeowners to buy wisely:
1) Shop around for the best service at a reasonable price.
2) If you do not know how to recognize insect infestation or damage, ask the salesman to show it to you. He must have evidence to recommend treatment.
3) Have the salesman certify in writing that the premise or structure is infested with a specific pest and that a treatment is required.
4) You may confirm the salesman's diagnosis by sending evidence of the pest or damage he shows you to the Cooperative Extension office in your county. An unbiased opinion may insure the proper treatment.
5) Get opinions from 2 or more pest control firms before deciding on expensive and extensive treatments.
6) Read the proposed contracts carefully so you know exactly what the company will do and guarantee. Ask the salesman to interpret parts that you do not understand.
7) Check with the Better Business Bureau to be certain the company has a clean record.

By following these procedures you are more likely to buy the services you need and want.

CONTRACTS FOR WOOD-DESTROYING ORGANISMS CONTROL

A pest control company must give you a written contract for each wood-destroying organism control or preventive treatment. The contract should be furnished before the work has begun and before any payment is made and contain the following information:

1) Name and address of property owner and address of property to be treated.
2) All buildings or structures to be included for treatment.
3) The complete name and address of the pest control firm.
4) Date of agreement, duration of contract, and renewal option (if offered).
5) Name of wood-destroying organisms covered by the contract.
6) Whether treatment is recommended for control of existing infestation or prevention of infestation.
7) Conditions under which the company will retreat in case of reinfestation.
8) Conditions under which repairs will be made.
9) Maximum price to be charged for treatment.
10) Whether reinspections are to be made and, if so, how often, and how much they will cost.
11) What renewal fees will cost.
12) Maximum price for structural repairs, if any, shown separately.
13) If guaranteed with a bond, what the obligations of the bond are; for instance, retreatments, repairs, etc.

Wood-destroying organisms do not destroy property overnight. Take your time to read the contract and compare services. For more information about wood-destroying organism control, write to:

Department of Agriculture
and Consumer Services
Bureau of Entomology & Pest Control
P. 0. Box 210
Jacksonville, Florida 32231

and ask for their free publication "Consumer Service Aid for the Homeowner and Buyer - Termites and other Wood-Infesting Pests".

WHAT TO LOOK FOR IN THE DELIVERY OF SERVICE

All company employees who come to your house reveal the professionalism of the firm. Look for the following:
1) Is the vehicle in which he arrived dirty and messy? Is the vehicle plainly marked with the name of the company with whom you are dealing?

2) While performing the job, is he considerate, courteous, and careful with your belongings?

3) Does he replace items he moves to do the job?

4) Is he organized or must he continually run to the truck for supplies?

If you are impressed with all the factors about the technician, you have reason to believe he may be conscientious in his performance of pest control. However, all these are superficial and not alone a true indication of the effectiveness of the services he has rendered.

There are several direct indications of the quality of pest control services the company is delivering. Companies that are concerned about your safety will follow several basic rules. These are:

1) Pesticides should never be diluted in your house. Pesticides should always be diluted in or at the service vehicle or at the company's business location. Any applicator who dilutes his pesticide in your home is not using approved pest control practices.

2) The technician should always leave his service equipment and pesticides in a locked vehicle or service compartment when not using them.

3) Pesticides should be applied indoors with low pressure to prevent splashing and runoff. Any spills or puddles should be wiped up immediately by the technician. He should come into your home with rags, etc., to wipe up accidental spills.

4) Every time a spray is applied indoors, you should be told to keep pets and children away from surfaces until they are dry. He should also request that anyone suffering from allergies, heart ailments, or respiratory problems be kept out of the room during the application, or that application be delayed until such persons can be removed.

5) Pesticides should never be applied to food products, utensils, or to surfaces that food will contact.

6) The applicator should be prepared to protect caged pets by covering them and should shut off air pumps in fish tanks. If cages are removed from rooms, he should check them to make sure they are free of pests.

7) The technician should never apply waterbased pesticides to electrical equipment.

8) You should be warned to keep children and pets off lawns until the chemical has dried.

9) Poison baits should be placed where children or pets cannot get to them.

10) The technician should point out any damage present before spraying, such as stains, burns, etc., so you will not think his spraying caused the problem.

11) Technicians should never give or sell pesticides to customers. Never ask the technician to give you a pesticide.

12) Children's toys, mattresses, and bedding should never be sprayed with pesticides.

COMPLAINTS ABOUT PEST CONTROL FIRMS

It is against the law for a pest control firm or its employees to perform pest control in a negligent manner. It is also against the law for any pest control firm or its employees to knowingly make false or fraudulent claims, misrepresent, the effects of materials or methods, or fail to use methods or materials suitable for the pest control undertaken. Fraudulent or misleading advertising is also against the law.

Pest control firms cannot tell you that a pest is infesting your property or structure, or tell you that a specific treatment for pest control is required, unless supporting evidence of an infestation exists.

A pest control firm violating the law may have its license or certificate suspended or revoked if the claims are proven. If you have a pest control complaint regarding any negligence, fraudulent claims, or improper use of materials and methods, contact the nearest Entomologist-Inspector with the Florida Department of Agriculture and Consumer Services.

Table I - Entomologist-Inspectors with the Florida Department of Health and Rehabilitative Services

Location	Phone Number
W. T. Edwards Bldg. 4000 W. Buffalo Avenue Room 234 Tampa, FL 33614	(813) 871-7020/1 SC: 594-2411
300 31st St. N, Ste. 524 St. Petersburg, FL 33713	(813) 893-2486 SC: 594-2411
401 N. W. 2nd Ave. South Tower, Rm. S-411 Miami, FL 33128	(305) 377-5968 -5082 SC: 452-5968
P.O. Box 1552 Marianna, FL 32446 (ask for Environ. Health)	(904) 526-2412 (904) 482-7203 SC: 775-1142
411 S. W. 4th Avenue Gainesville, FL 32601	(904) 336-2270/1 SC: 625-2270/1
P.O. Box 29 West Palm Beach, FL 33402	(407) 837-5210 SC: 252-5210
Hurston Tower South 400 W. Robinson St., S-357 Orlando, FL 32801	(407) 423-6837 SC: 344-6837
State-wide Mosquito Control	(407) 423-6839 SC: 344-6839

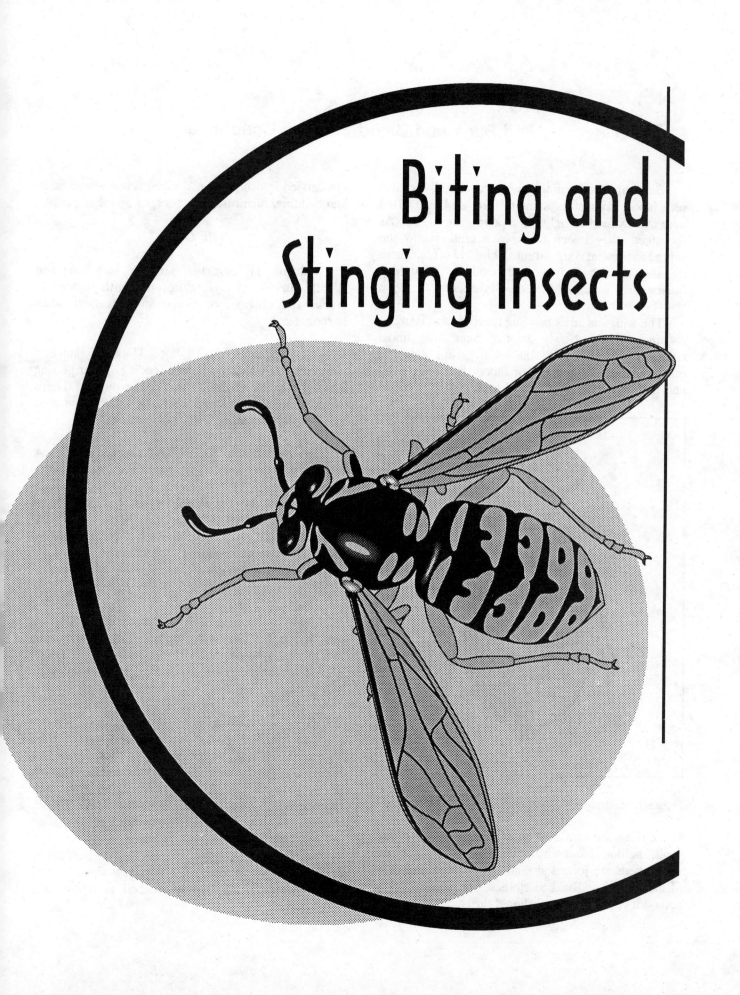

Biting and Stinging Insects

20

Bed Bugs and Blood-Sucking Conenose

Bed bugs feed mainly on human blood, but also suck blood from other animals, birds and bats. Bed bugs usually feed at night when people are asleep. As they feed, they inject a salivary secretion into the wound to prevent coagulation. This fluid often causes the skin to itch and become swollen. Scratching causes sores which may become infected.

The wingless adult bed bug is brown and is about 1/4 inch long. Newly hatched bugs are almost colorless and similar to the adult except they are much smaller. When full of blood, bed bugs swell and their color changes to dark red.

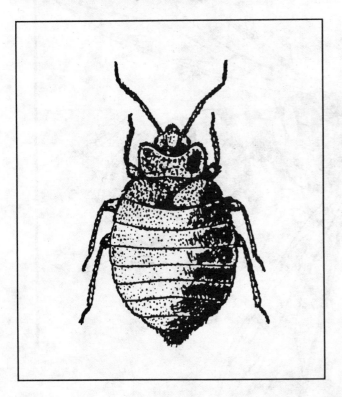

Figure 1. Bedbug.

During the day, bed bugs hide in cracks in the walls, behind baseboards, wallpaper and pictures, where beds are joined around slats, around the tufts of mattresses and in bed clothes. The bugs have a bad odor caused by an oily liquid they emit. Bed bugs

are carried into homes in clothes, second-hand beds, and bedding, furniture, suitcases, or by other people.

Life Cycle

Bed bugs lay eggs that are 1/25 inch long and slightly curved. They are fastened with cement to cracks and crevices or rough surfaces near adult harborages.

The eggs hatch in 6-10 days. The newly hatched nymph is straw colored before feeding. After getting a blood meal, the nymph turns red or purple. There are 5 nymphal stages and it usually takes 35-48 days for nymphs to mature.

Female bed bugs deposit 10-50 eggs at a time. A total of 200-500 eggs can be produced per female. The eggs hatch in 1-3 weeks. Adult bed bugs can survive for 6-7 months without a blood meal and have been known to live in abandoned houses for 1 1/2 years. In some cases they survive without humans by attacking birds and rodents.

Control

Steam cleaning of infested mattresses kills bed bugs living in seams and buttons. Cracks harboring bed bugs can be treated with residual sprays. Total-release aerosols can also be used to treat rooms with bed bugs.

Take the bed apart. Spray the bed frames, slats and springs with enough spray to thoroughly wet them. Pay particular attention to the tufts and seams of the mattress. Spray the woodwork and all walls in the bedroom at least 2 feet above the floor. Brush, vacuum, and steam clean mattress and pillows, then put on clean sheets and pillow cases.

Spray again if there are any new signs of bed bugs. After 2 weeks, spray the bed, furniture, and walls again.

BLOOD-SUCKING CONENOSE

The blood-sucking conenose is a brown, winged bug, 3/4 inch long with the edges of its abdomen alternating light and dark color. It has piercing-sucking mouthparts. The conenose's life cycle varies considerably depending on availability of hosts. There are usually 1 to 2 generations per year.

This ectoparasite feeds on blood of sleeping persons at night. It is found in bedding, cracks in floors and walls, and under furniture. The blood-sucking conenose is a potential vector of Chagas disease.

Control

Apply residual sprays to suspected hiding areas.

Non-chemical control may be obtained by screening windows, since conenoses are attracted to lights at night. Eliminate breeding areas.

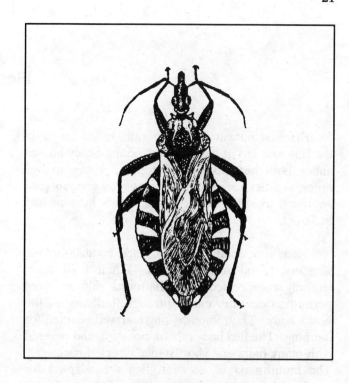

Figure 2. Blood-sucking conenose.

22

Fleas

External parasites are generally found on or in the skin and are important pests because they bite or annoy both humans and their pets. Fleas, mange mites, and ticks are the most frequently encountered and most troublesome pests that attack humans and their pets.

Fleas are small (1/16 in.), dark, reddish-brown, wingless, blood-sucking insects. Their bodies are laterally compressed, (i.e., flattened side to side) permitting easy movement through the hairs on the host's body. Their legs are long and well adapted for jumping. The flea body is hard, polished, and covered with many hairs and short spines directed backward. The mouthparts of an adult flea are adapted for sucking blood from a host.

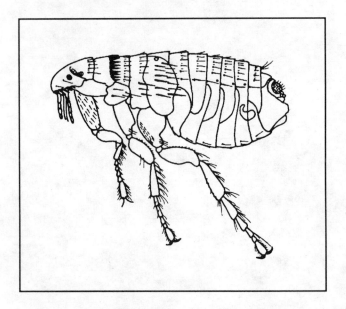

Figure 1. Flea adult.

Several species of fleas may be pests in Florida, and five kinds have been found on a single animal. The cat flea is the most frequently found, although the dog, human, and sticktight fleas are also found in Florida. Fleas may attack a wide variety of warm-blooded animals including dogs, humans, chickens, rabbits, squirrels, rats and mice.

BIOLOGY

The female flea lays her tiny, white eggs loosely on the hairs, in the feathers, or in the habitat of the host. The eggs readily fall off the host onto the ground, floors, bedding, or furniture. Some fleas can lay 500 eggs over a period of several months by laying batches of 3-18 eggs at a time. The tiny eggs hatch in 1-12 days after being deposited. The white, worm-like larva (Figure 2) avoids light and feeds on particles of dead animal or vegetable matter generally present in cracks and crevices. Within 7-14 days, unless food has been scarce, the third larval stage is completed, and the larva spins a tiny cocoon and pupates. Usually after a week the adult flea emerges and begins its search for blood.

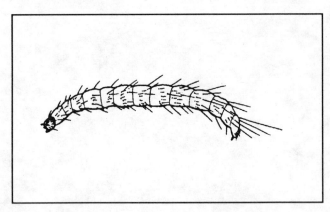

Figure 2. Flea larva.

Fleas are known to remain in the pupal stage from 5 days to 5 weeks in the absence of hosts. Adults emerge from the pupal case when vibrations from pets or humans let them know a host is near. This is one reason why people returning to an unoccupied home may suddenly be attacked by an army of fleas.

Adult fleas must feed on blood in order to reproduce; however, adults can live for long periods without feeding. Fleas usually live and breed most heavily where pets rest. Persons coming near these resting places are also subject to attack. If fleas are established in a home, they will feed on man as well as on the pets. The usual places of attack are the ankles and lower portions of the legs.

The so-called "sand-flea" is nothing more than a common flea that is breeding outdoors in the soil. Contrary to belief, fleas cannot go through several generations without having a blood meal.

The entire life cycle of a flea requires from 2 weeks to 2 years. Hot, wet, summer months favor egg laying. Hot, dry periods give maximum adult production, so greatest adult flea populations are produced in August to September.

IMPORTANCE

Fleas often breed in large numbers where pets and other animals live. Pets infested with fleas bite and scratch themselves constantly. Their coats become roughened and the skin can become infected. Symptoms of sensitized hosts are often mistaken for mange. Cat fleas and dog fleas may be intermediate hosts for the dog tapeworm.

Some people suffer more from flea bites, because these can cause intense itching often resulting in secondary infection. The usual flea bite has a small red spot where the flea has inserted its mouthparts. Around the spot there is a red halo with very little swelling. Many people do not react to flea bites at all while others are sensitive and suffer severe allergic reactions. Fleas may also vector such human diseases as plague, typhus, and tularemia.

CONTROL

Flea control is difficult for pet owners to implement because two things must be done: (1) treat the pet and (2) treat the premises. Pet treatment alone is not sufficient because the animal quickly becomes reinfested from untreated premises.

PET TREATMENT

Flea collars are sold under several trade names and are sometimes effective on small, short-haired dogs or cats that are not subjected to flea-infested areas.

Other treatments are usually necessary to supplement flea collars on large, long-haired pets that are allowed freedom outdoors. Also, some pets may be allergic to flea collars. Ultrasonic flea collars have not been found to kill or repel fleas.

Veterinarians may prescribe or apply pesticides not available over the counter. Oral flea medication prescribed by veterinarians has provided control of fleas when pets are not allowed outdoors and effective flea control is accomplished in the house and yard.

Pets may be combed or shampooed frequently to remove adult fleas before they can irritate the pet or lay eggs. Frequent removal of fleas can quite effectively reduce flea infestations.

To be certain pets remain free of fleas, it is necessary to make routine use of flea control products, especially if pets are allowed to contact infested animals or premises.

Dust treatments should be applied carefully and rubbed into the fur working from the head to the tail. Special attention should be given to the top of the head, the neck, and the back. Apply treatments outdoors so fleas that leave the animals do not remain indoors.

PREMISE TREATMENT

Pets become reinfested with fleas from premises. For the most effective control, sleeping areas, bedding kennels, and other areas frequented by the animal should be treated at the time the pet treatment is made. Treatments may or may not include the use of pesticides.

Nonpesticidal premise control includes thorough and frequent cleaning of the house. All rugs should be thoroughly cleaned with vacuum cleaner or steam cleaner. Infested furniture, pet baskets, and cracks should be thoroughly cleaned to prevent the larvae from finding food. Dirt which is collected should be disposed of immediately to destroy fleas and flea larvae.

Many people remove pets from the home to attempt flea control. Flea infestations usually become more evident when pets are removed. The hungry adult fleas prefer to feed on cats and dogs. When the pet is removed, the fleas overrun the home, frequently attacking humans. Dogs and cats can be used to attract fleas from the premises. Recommended pet treatments at frequent intervals can be used to kill the fleas.

Spot treat with insecticides when fleas become established on a pet or in the home. Apply sprays according to label directions and do not apply directly to pets. Sprays are effective when properly applied to surfaces in the house. Insecticides can be applied for yard treatment. Sprays should be directed to all known or suspected breeding places.

Insect growth regulators (IGRs) are the most effective of chemicals and are found in indoor misters and total release aerosols. These may be used in conjunction with residual sprays to quickly reduce adult populations. Be sure to follow the label directions. IGRs prevent flea larvae from turning into adults, and have a residual effect of almost three months. For IGR applications to be effective, pets must not be allowed access to heavily infested areas outdoors; otherwise adult fleas will constantly be carried indoors by the pet.

Reduced Chemical Control Of Fleas

NON-CHEMICAL METHODS OF REDUCING FLEA POPULATIONS

Cat fleas (*Ctenocephalides felis*) are the most common fleas found on both cats and dogs. Flea eggs and adult feces fall off the host and accumulate where the pet rests, making these the prime flea breeding foci. Flea breeding in these areas can be reduced by establishing one sleeping area for the pet, choosing an area that can be cleaned thoroughly on a regular basis, and using bedding material that can be laundered weekly or thrown away as often. If human flea allergy is the cause of the complaint, pets should be excluded from areas of the house frequented by the person(s) afflicted with the allergy.

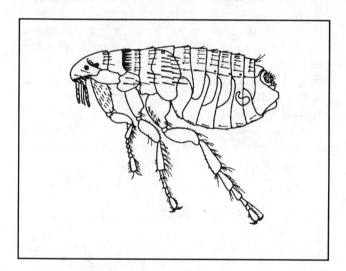

Figure 1. Adult cat flea.

Vacuuming flea breeding areas is also beneficial. Byron and Robinson (1986) found that a beater-bar vacuum would remove 15 - 27% of larvae introduced into a carpet and 32 - 59 % of the eggs. In addition to removing the juvenile stages, vacuuming will also remove adult flea feces, the essential food source for the developing larvae. Pupae appear to be unaffected by vacuuming, as their silk cocoons are tightly bound to carpet fibers. Attachments should be used to vacuum cracks, crevices, and upholstered furniture where pets rest. Vacuuming alone will not remove the entire flea population but will help keep them reduced. The steam-extraction carpet cleaning method should effectively destroy all stages of fleas present.

Combing the pet with a flea comb is an effective but time-consuming method of controlling fleas on pets. It is most effective on cats, especially since cats do not normally tolerate baths. Fleas removed with the comb should be disposed of by dropping them in soapy water (Olkowski et al. 1983). Our experiences have shown that many cats find combing enjoyable. (Especially if you give the cat treats afterwards!)

Bathing a dog with soap, shampoo or a mild detergent can also remove fleas. This method drowns fleas (Olkowski et al. 1983) and the lipophilic (oil attracting) nature of detergents should have a drying effect on the surviving fleas. Olkowski et al. (1983) recommended the use of insecticidal soaps such as "Safer Flea Soap for Cats and Dogs" (25% potassium salts of fatty acids).

Flea traps will capture some fleas, but there is no evidence that they do anything to control flea populations. They may have some small value as a sampling tool to confirm infestations. Ultrasonic pest repelling devices have repeatedly been shown to be completely useless (Hinkle et al. 1990 , Dryden et al. 1989). The use of leaves from waxmyrtles, such as the southern bayberry (*Myrica cerifera*), have been repeatedly recommended for flea control. Tests on packets of dried bayberry leaves showed no repellent effect on cat fleas (W. H. Kern & R. S. Patterson, unpub. data). The effectiveness of fresh bayberry or other repellent plant materials, such as penny royal, eucalyptus, rosemary, or citronella has not been documented (Olkowski et al. 1983).

BIORATIONAL CHEMICALS

Insect Growth Regulators

IGRs are the most effective weapon in the flea control arsenal. They are effective at very low concentrations (< 10 ppb) and have virtually no mammalian toxicity. They mimic insect hormones and act on insects by interfering with their normal development. If you can tolerate fleas for the time it takes these compounds to eliminate a flea population, IGRs are effective by themselves. Expect to wait 2 weeks before noticeable flea reductions and 1-2 months before complete control.

Conventional Pesticides

People have different levels of acceptance to pesticides. Some may accept pyrethrum because it is a botanical, an extract of dried chrysanthemum flowers. But those with hay fever, especially allergy to ragweed, may show cross reactivity to pyrethrum making it unacceptable. Pyrethrum is known for its rapid knockdown, but fleas often revive and recover in time. Many people perceive boric acid or borax as being non-toxic and it is very effective for treating carpets.

On-Host Treatments

It is necessary to treat the pet at the same time as the premises. Dusts are considered by some to be safer than sprays or dips (Olkowski et al. 1983).

Flea collars are the most seriously abused method of flea control. The advantage of a flea collar is that it can be placed on an infested pet for a treatment period (optimally less than 6 days) then removed and stored in a sealed glass jar until the next time it is needed. Flea collars should not be kept on pets permanently as a prophylactic measure. Be sure to check for dermatitis under the collar (Olkowski et al. 1983).

Dusts are considered safer than sprays or shampoos since there is no solvent to carry the pesticide through the host's skin. The problems with dusts or powders are that can't be controlled, easily becoming air-borne and they are ingested by cats during grooming. It is recommended that a pet be bathed after being treated with a pyrethrum dust to remove the fleas stupefied by the pyrethrum and reduce the risk of accidental ingestion by the pet.

Repellents can be used if you are taking your pet into a known flea-infested area. If a flea population exists in the home, use of a repellent on the pet means that the fleas will switch to their second choice

for a blood meal - YOU. Pet repellents containing DEET are available. Many people report that the skin softener, Avon's SkinSoSoft, helped reduce flea problems for their pets. This product is not registered as an insecticide or for flea control.

Premises Treatment

Carpets can be effectively treated with boric acid or borax, alone or in combination with diatomaceous earth. Boric acid should be applied as a dust to indoor flea breeding areas. Boric acid may eventually cause damage to carpet fabric or upholstery. Boron compounds should not be used outdoors since they act as non-selective herbicides and will kill most plants.

Diatomaceous earth is made up of the fossil shells of single celled algae called diatoms. The shells are chemically and physically like ground-up glass. Diatomaceous earth kills only flea larvae. The larvae are scratched by sharp edges, lose body moisture, and die from desiccation. It has been found to be effective in dry climates but is much less effective in humid Florida. Linalool is an extract from citrus peels which is registered for indoor flea control. A wide range of conventional insecticides are registered for flea control as water diluted sprays, total release aerosols, or hand-held aerosols. Most of these would be unacceptable to pet owners who are uncomfortable with synthetic insecticides. No matter what product is used, always follow label directions.

REFERENCES CITED

Byron, D. and W. Robinson. 1986. Much ado about fleas. Pest Control 1986(May): 40-42, 47.

Dryden, M. W., G. R. Long, and M. G. Sayed. 1989. Effects of ultrasonic flea collars on *Ctenocephalides felis* on cats. J. Am. Vet. Med. Assoc. 195(12): 1717-1718.

Hinkle, N. C., P. G. Koehler, and R. S. Patterson. 1990. Egg production, larval development, and adult longevity of cat fleas (Siphonaptera: Pulicidae) exposed to ultrasound. J. Econ. Entomol. 83(6): 2306-2309

Olkowski, W., H. Olkowski, and S. Daar. 1983. IPM for the cat flea. The IPM Practitioner 5(9): 7-11.

Human Lice

Three kinds of lice live and breed on man: head, body, and crab lice. All can live on the human body and suck blood. Pediculosis or lousiness is one of the most prevalent communicable conditions in this country. Lice are transferred from person to person by direct contact or by several people using the same combs, brushes, or bedding. Human lice are not found on animals or household pets and are not transmitted from pets to humans.

HEAD LOUSE AND BODY LOUSE

Head louse infestations are more prevalent than those of the body louse. Head louse infestations are normally found on children, but can also be spread to adults. The head louse is not considered to be a serious vector of disease although severe infestations may cause irritation, scratching and subsequent invasion of secondary infection.

Body louse infestations are normally associated with poor personal hygiene and poverty. Body lice are capable of transmitting louse-borne typhus, but in the United States the disease is not prevalent.

Biology

Head lice prefer to live on the hair of the head although they have been known to wander to other parts of the body. The body louse prefers to remain on the clothing of the host and feed on the body. A person infested with hundreds of body lice may remove his/her clothes and not find a single louse on his/her body.

The eggs of lice are called nits. They are oval white cylinders. The eggs of head lice are usually glued to hairs of the head. The favorite areas for females to glue the eggs are near the ears and back of the head. Body lice glue their eggs to clothing, especially near seams and creases.

Female lice lay 6-7 eggs per day and may lay a total of 50-300 eggs during their lives. Under normal conditions the eggs will hatch in 5-10 days, averaging seven days. The young lice that escape from the egg

must feed within 24 hours or they will die. The nymphs and adults all have piercing-sucking mouthparts, which pierce the skin for a blood meal. Adults may survive 3-5 days without a blood meal. Normally a young louse will mature to an adult in 3-5 weeks.

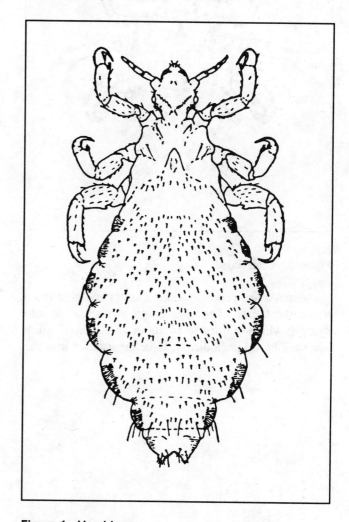

Figure 1. Head louse.

Bites

The reaction of individuals to louse bites can vary considerably. Persons previously unexposed to lice experience little irritation from their first bite. After a short time individuals may become sensitized to the bites and may have a general allergic reaction.

28

CRAB LOUSE

The crab louse has a short broad body and the general appearance of a crab. It is well adapted for survival where coarse hair grows, especially in the pubic regions. Crab lice don't move around as much as head and body lice. They are usually transferred by direct contact.

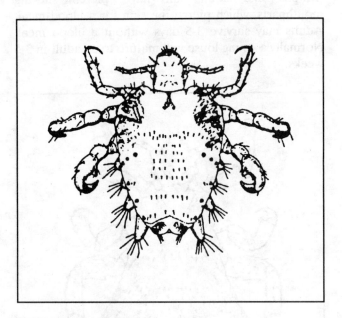

Figure 2. Crab louse.

Female lice produce about 25 eggs and glue them to coarse hairs. The eggs hatch in one week into nymphs which mature to adults in 15-20 days. Adult lice suck blood periodically, injecting a saliva into the wound which causes pale blue spots from 0.125-1.0 inch in diameter.

Control of Lice

Body lice may be controlled by changing clothes daily. All clothes should be washed in hot soapy water. Pillow cases, sheets and blankets should also be washed to kill the lice and their eggs.

Head lice should not be solely associated with uncleanliness since they may be easily transferred from person to person.

Chemicals are available as prescription or non-prescription drugs to control lice. These drugs are available as creams, lotions, or shampoos. Shampoos are preferred for control of head lice. The application of these insecticidal drugs will kill nymphs, adults and some eggs. Because some eggs may survive, it may be necessary to treat once per week for three weeks to insure complete control. Even though the nits may be killed, the unsightly eggs may remain glued to the hair. These eggs may be removed by combing with a fine-toothed comb or nit picking.

Because nymphs and adults sometimes fall off the host, areas such as cubby holes in schools, lockers and other places may be treated. These treatments should be carefully applied to surfaces not directly contacted by children. Several aerosol sprays can be used to treat infested beds or furniture. Apply the sprays as directed on the label.

Mites That Attack Humans

Mites are small arthropods with two body regions, sucking mouthparts, no antennae, and four pairs of legs as adults. The life cycle of a mite is generally composed of four active stages: egg, larva, nymph, and adult. The life cycle usually requires one to four weeks and may result in huge populations of mites when there are favorable conditions.

HOUSEHOLD MITES

Mites are occasionally found in homes and attack humans in the absence of their normal hosts - birds, rodents, or insects. Bites from these mites may be painful and cause severe skin irritation.

BIRD MITES

The northern fowl, tropical fowl, and chicken mites are the major bird-mite species in Florida. The northern fowl mite is the species most important as a problem to birds in the state.

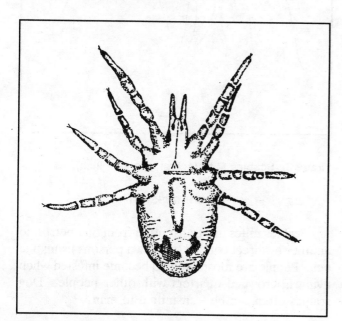

Figure 1. Northern Fowl Mite (actual size 1/16 inch).

The adult female fowl mite lays eggs on the host bird. The eggs hatch in 1-2 days into the six-legged larval stage which does not feed. The larvae molt to

the nymphal stage in about eight hours. The nymphs and adults have piercing mouthparts and seek blood meals. The complete life cycle from egg to egg-laying adult can take from five to seven days or longer, depending on the environment.

Bird mites are usually encountered in homes when they migrate from bird nests in eaves, rafters or gutters. They prefer to feed on fledglings in the nest, but when the young leave the nest, the mites will migrate to other areas in search of a blood meal. Many times infestation of buildings occurs when roosts and nests of birds are disrupted or destroyed.

INSECT MITES

The almost-invisible straw itch mite is the most prevalent insect parasite that also attacks humans. Infestations from alfalfa, hay and barley can produce irritation.

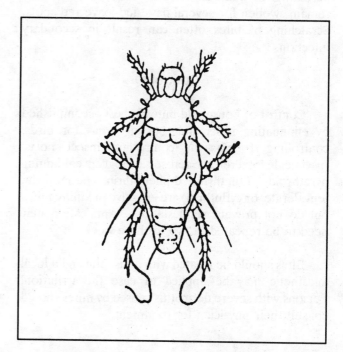

Figure 2. Straw itch mite (actual size 1/32 inch).

The mites are parasitic on the larvae of insects such as the angoumois grain moth, the wheat jointworm, and furniture beetles. The female mite retains up to 300 eggs in her body, and the immature stages develop to adults inside the female. Upon emerging they search for hosts to parasitize.

The bites of the straw itch mite are located almost entirely on the clothed portions of the body. Dermatitis results from reaction to the bites within twenty-four hours. Humans become infested when they come in contact with straw, grain or wood. Houses may become infested when the insect hosts of the mites are present.

RODENT MITES

The two most abundant rodent mites in buildings are the tropical rat mite and the house mouse mite. Rodent mites are primarily external parasites of rats and house mice, but they will also feed on humans.

The life cycle of rodent mites is similar to that of the bird mites. The life cycle usually takes from 10 to 12 days.

Rodent mites can cause severe irritation and dermatitis in humans. Areas bitten by mites may remain swollen for several days and leave red spots. Scratching of bites often can result in secondary infection.

Control

Control of household mites is best accomplished by eliminating nests and roosting areas for birds, controlling rodents or controlling insect hosts. Insecticide total release aerosols or foggers containing pyrethroids (allethrin, d-phenothrin, permethrin, fenvalerate, or cyfluthrin) are effective in killing mites but do not prevent reinfestation. Application may need to be repeated in two to three weeks.

Bites should be treated with antiseptic and a local anesthetic may be applied to ease the irritation. Persons with severe dermatitis caused by mites should consult their physician for treatment.

SCABIES MITES

Biology

The scabies mite or human itch mite burrows into the skin of humans causing human mange or scabies. Different varieties of scabies mites are specific for certain mammals including man, domestic animals, and wild animals.

The female mite makes long burrows in folds of skin. The female lays from 40 to 50 eggs in the burrow. The larvae and nymphs develop and burrow in the skin. The total life cycle takes one to three weeks depending on the environment.

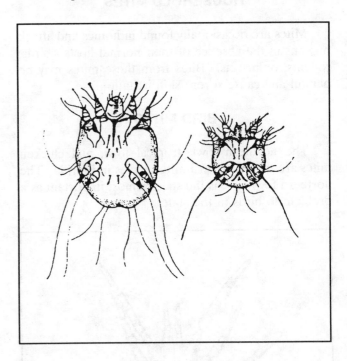

Figure 3. Scabies Mite (actual size 1/64 inch).

Transmission

Scabies mites are transmitted from one person to another by direct contact or by two persons using one bed. People are most likely to become infested when living in crowded quarters with other people. Dog scabies often can be transmitted to man.

Symptoms

Persons infested with scabies suffer severe itching. A rash may develop later as the person becomes sensitized to the mites. The rash usually occurs around armpits, the wrists, the waist, and back of the calves. Even though only a few mites may be present, the rash may spread over much of the body.

Control

If scabies mites are suspected, a physician should be consulted for diagnosis and treatment. If dog scabies transmission to man is suspected, a veterinarian should be consulted for treatment of the dog.

HOUSE DUST MITES

The house dust mites (floor mites) are a little known and rarely mentioned group of mites of medical significance to humans. The importance of house dust mites is that pieces of the mites in house dust may produce allergic reactions when inhaled. It is presently estimated that four percent of the human population shows a house-dust allergy.

Biology

The adult, female, house dust mite is approximately 1/64 inch long and the males are even smaller. Because of their small size these mites are often overlooked.

The adult female lays about an egg a day for 30 days. The eggs hatch and develop into adults in about one month.

House dust mites feed on the shed skin of humans. It has been estimated that a normal person sheds five grams of skin per week. One gram of skin will feed thousands of mites for months.

The house dust mite is found commonly in houses and schools throughout the United States. It is believed that dust near the bed and in mattresses is the most common habitat for the mite.

Mite Allergy

House dust mites or their fragments, excretory or secretory products are the most important allergens found in house dust. Asthmatic symptoms after exposure to house dust have long been known and only recently has it been shown that the house dust mites, like pollen, can be the potential producers of allergic responses when inhaled.

Asthmatic patients who go to bed in the bedroom often realize aggravated asthmatic symptoms due to increased exposure to house dust mites. Patients who are hospitalized often dramatically improve due to less exposure.

Control

No chemical control of house dust mites is presently known. Frequent changing of bedding and use of non-fibrous bedding will reduce mite populations. Frequent vacuum cleaning and correction of excess humidity problems will aid in mite control.

Chiggers

Chiggers or "red bugs" are the larvae of mites belonging to the family Trombiculidae. In humans, chiggers can cause intense itching and small reddish welts on the skin. In other parts of the world, chiggers transmit scrub typhus; however, in Florida they are not known to transmit disease. The intense irritation and subsequent scratching may result in secondary infection.

Biology

Adult chiggers usually overwinter in protected places and become active in the spring, although in Florida they may be active all year. The females lay eggs in a sheltered area. The eggs hatch into an ectoparasitic, six-legged larval form which is less than 1/50 inch long. The larva is the parasitic stage that feeds on humans and animals. The orange-yellow or light-red larval stage crawls on the soil surface until a suitable host is found. Suitable hosts range from mammals, including humans, to birds, reptiles, and amphibians.

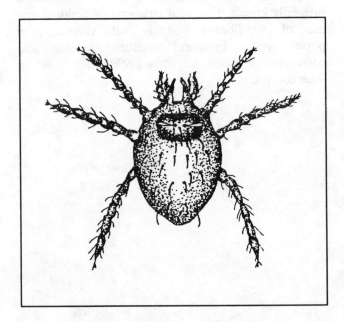

Figure 1. Chigger (actual size 1/64 inch).

The larvae suck fluids from the skin of the host animal and do not burrow in the skin. Usually within three days the larva is engorged. The larva then drops from the host to the ground and transforms into the nymphal stage.

The nymph, like the adult mite, has eight legs. The bodies are usually hairy and about 1/20 inch long and bright red. The nymphs and adults feed on insect eggs, small insects, or other organisms.

The entire life cycle can require from two months to one year. As many as 1-5 generations may be produced per year depending on the temperature, moisture, and location.

Chigger Bites

Chiggers attach themselves to the skin, hair follicles or pores by inserting their piercing mouthparts. When chiggers attach to humans, they are not usually noticed for some time. During feeding, they inject a fluid into the skin which dissolves tissue. Chiggers feed by sucking up the liquified tissues.

Itching from chigger bites is usually noticed 4-8 hours after chiggers have attached or have been accidentally removed. The fluid injection causes welts to appear which may last for two weeks. Some people exhibit an allergic reaction to the injected fluid which results in severe swelling, itching, and fever. People mistakenly believe that chiggers embed themselves in the skin or that the welts contain chiggers. Often scratching at the welt results in secondary infection.

Chiggers prefer to attach on parts of the body where clothing fits tightly or where the flesh is thin, tender, or wrinkled. For this reason, chiggers locate in such areas as the ankles, waistline, knees, or armpits.

Preferred Habitat

Chiggers are most often found in low, damp areas where vegetation is heavy, although some species prefer dry areas. Chiggers seem to be most abundant in areas covered with shrubs and small trees where rodents are numerous. Chiggers occur in pockets or islands because a female will lay all her eggs in one spot; however, chiggers may persist in home lawns.

Control

Chiggers are easily removed from the skin by taking a hot bath or shower and lathering with soap several times. The bath will kill attached chiggers and others which are not attached. Since symptoms of contact may not appear for several hours, it is not always possible to completely prevent welts caused by chigger bites. Antiseptic should be applied to all welts which do appear. Temporary relief of itching may be achieved with nonprescription local anesthetics available at most drug stores. Studies have shown that meat tenderizer, rubbed into the welt, will alleviate itching.

If you are going into areas suspected of being infested with chiggers, wear protective clothing and use repellents. Repellents should be applied to legs, ankles, cuffs, waist, and sleeves by clothing application or directly to the body as directed by the label.

Chigger infestations may be eliminated by insecticide applications or by clearing brush from the area. To locate infested areas, place a piece of black cardboard edgewise on the ground. If chiggers are present, they will climb to the top and congregate there. They will appear as tiny yellow or pink dots moving across cardboard. Before starting your survey for chiggers, use protective clothing and repellents.

Area control of chiggers is difficult to achieve. The best method is to mow grass close to the ground. Mowing removes the protective cover that chiggers need to survive. Insecticides may be applied for area control of chiggers. Apply these chemicals as directed on the label, and avoid getting insecticides into ponds and streams.

Poisonous Spiders

Most spider bites are not likely to be dangerous, but medical care and advice should be sought in each case and it is important to save any biting spider so it can be identified.

WIDOW SPIDERS

Four species of widow spiders occur in Florida: the southern black widow, the northern black widow, the red widow and the brown widow. All these species are rather large spiders, about 1 1/2 inches long with the legs extended.

The southern black widow and the northern black widow are a shiny, jet-black color. The southern black widow has a red hourglass marking on the underside of the abdomen and another red spot at the tip end of the abdomen (Figure 1). The northern black widow has a row of red spots located in the middle of its back and two reddish triangles resembling an hourglass on the underside of the abdomen. The red widow spider has a reddish orange head-thorax and legs with a black abdomen. The abdomen may have a dorsal row of red spots with a yellow border. The red widow lacks a complete hourglass under the abdomen but may have one or two red spots. The brown widow spider varies in color from gray to light brown or black. The abdomen has variable markings of black, white, red, and yellow. On the underside of the abdomen the bown widow has an orange or yellowish-red hourglass marking.

The life cycle of the widow spiders are all similar. The female lays approximately 250 eggs in an egg sac which is about 1/2 to 5/8 inch in diameter. The eggs hatch in 20 days and remain in the egg sac from about 4 days to 1 month. The young spiders then molt to the second stage and begin feeding. As the young spiders grow, they construct a loosely woven web and capture progressively larger prey. Male spiders molt 3 to 6 times before maturing. The females molt 6 to 8 times and occasionally eat the males after mating. In Florida all the widows except the northern black widow breed year-round.

Figure 1. Black widow spider (ventral view, actual size 1 1/2").

The southern black widow is the most widespread widow spider in Florida. It is usually found outdoors in protected places such as in hollows of stumps, discarded building materials, rodent burrows, storm sewers, and under park benches and tables. Around houses, the southern black widow is found in garages, storage sheds, crawl spaces under buildings, furniture, ventilators, and rain spouts. The northern black widow is found west of Tallahassee. It is mainly found in forests in irregular, loosely woven webs 3-20 feet above the ground. The red widow spider makes its web off the ground in palmetto habitats and has only been found in sand-pine scrub associations. The web retreat is characterized by the rolled palmetto frond, and the web is spread over the fronds. The brown widow is found only in coastal cities located south of Daytona Beach where it usually lives on buildings in well-lighted areas.

Like most spiders, the widow spiders are shy and will not bite unless aggravated. All four species have a strong venom. The southern black widow is involved in most poisonous spider-bite cases in Florida. The bite of the black widow is not always felt, but usually feels like a pin prick. The initial pain disappears rapidly leaving a local swelling where two tiny red spots appear. Muscular cramps in the shoulder, thigh, and back usually begin within 15 minutes to 3 hours. In severe cases, later pain spreads to the abdomen, the blood pressure rises, there is nausea and profuse sweating, and difficulty breathing. Death may result from the venom, depending on the victim's physical condition, age, and location of the bite. However, death seldom occurs if a physician is consulted and treatment is prompt.

If you suspect that a widow spider has bitten you, capture the specimen for identification and immediately consult a physician. For additional information, your doctor may wish to contact your local poison control center.

BROWN RECLUSE SPIDER

The brown recluse spider (Figure 2) is not an established species in Florida, but physicians have diagnosed its bites. The brown recluse spider is recognized by having a dark violin-shaped mark located behind the eyes. It has 3 pairs of eyes while most spiders have 4 pairs. The brown recluse is a medium-sized spider about 1/4 to 1/2 inch in length.

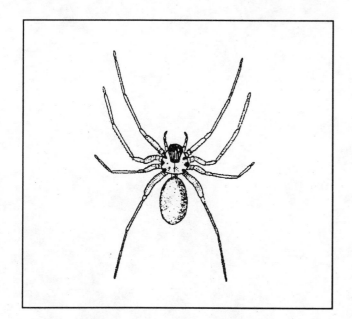

Figure 2. Brown recluse spider (actual size 1/4"-1/2").

The brown recluse spider's natural habitat is along the Mississippi River valley, especially in northwestern Arkansas and southern Missouri. Because it can live in old boxes and furniture it is easily transported by humans. Specimens of brown recluse spiders have been found in Florida, but there is no indication that it is able to survive and reproduce in Florida's environment.

The brown recluse spider is a shy species that bites humans when trapped in clothing or rolled onto when people sleep in bed. Persons bitten by the brown recluse usually do not feel pain for 2-3 hours. A sensitive person may feel pain immediately. A blister arises around the area of the bite. The local pain becomes intense with the wound sloughing tissue often down to the bone. Healing takes place slowly and may take 6 to 8 weeks. If the bite of a brown recluse spider is suspected, collect the spider and consult a physician immediately.

TARANTULAS

Tarantulas (Figure 3) are not found naturally in Florida; however, some people keep tarantulas as pets. The term "tarantula" refers to about 300 species of spiders some of which can weigh 2 to 3 ounces and have a 10-inch leg span. Tarantulas are sluggish, will not bite unless provoked, and are not poisonous. However, the bites of tarantulas can be quite painful since the fangs are large and can pierce the skin of the victim.

Many tarantulas have a dense covering of stinging hairs on the abdomen to protect them from enemies. These hairs can cause skin irritation for humans. Most tarantulas that are desirable as pets have a bald spot on the abdomen and do not have stinging hairs.

Tarantulas usually live in burrows in the ground. These burrows may be dug by the spider or abandoned by rodents. The tunnels are lined with silk and form a webbed rim at the entrance that conceals it. The females deposit 500 to 1000 eggs in a silken egg sac and guard it for 6 to 7 weeks. The young spiders remain in the burrow for some time after hatching and then disperse by crawling in all directions. Tarantulas do not occur in colonies because they do eat each other.

Tarantulas may live for many years. Most species require 10 years to mature to adults. Females kept in captivity have been known to live more than 25 years and have survived on water alone for 2 1/2 years.

36

Females continue to molt after reaching maturity and, therefore, are able to regenerate lost legs. Males live for only one year or less after maturity.

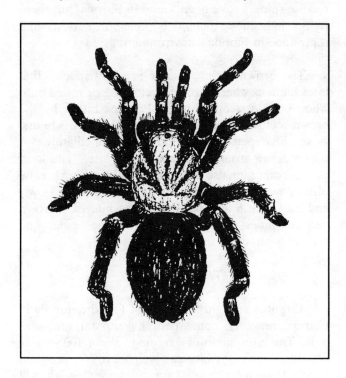

Figure 3. Tarantula (actual size up to 10").

A tarantula can be kept as a house pet. A terrarium (an empty aquarium) with a sandy bottom provides an ideal habitat. Tarantulas can be fed live crickets or other insects.

Mosquitoes And Other Biting Flies

MOSQUITOES

Mosquitoes can be an annoying, serious problem in man's domain throughout Florida. They interfere with work and spoil hours of leisure time. Their attacks on farm animals can cause loss of weight and decreased milk production. Some mosquitoes are capable of transmitting diseases such as malaria, yellow fever and dengue to man, encephalitis to man and horses, and heartworm to dogs.

Mosquitoes are insects with long slender bodies, narrow wings with a fringe of scales on the edge of the wing and along the veins, and long, thin legs. The females have firm mouthparts, usually well-adapted for piercing skin and sucking blood. The males cannot suck blood but both sexes feed on nectar of various plants.

The life cycle of a mosquito consists of four stages: egg, larva, pupa, and adult. The eggs may be laid singly or in rafts, deposited in water, on the sides of containers where water will soon cover, or on damp soil where they can hatch when flooded by rainwater or high tides.

Around the home suitable places for egg-laying are the sides of containers, such as old tin cans or old tires, or in tree holes to await flooding by rain. The eggs of some flood-water and salt-marsh mosquitoes may dry out for more than a year and still hatch when flooded.

Regardless of the mosquito species, water is essential for breeding. Mosquito larvae are not adapted to life in moving waters. The larvae normally occur in quiet water. Since nearly half the total land area of Florida is subject to flooding, mosquitoes breed in large numbers throughout the state. Contrary to popular opinion, mosquitoes do not breed in the heavy undergrowth of weeds, bushes, or shrubs. Although these places provide excellent refuge for adults, they do not provide a suitable habitat for mosquito larvae.

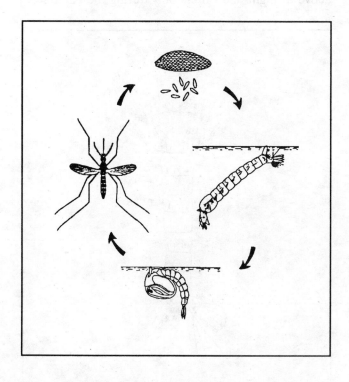

Figure 1. *Aedes aegypti* mosquito life cycle (eggs-top, larva-right, pupa-bottom, adult-left).

The elongated eggs, about 1/40 inch long, are laid in batches of 50 to 200 and one female may lay several batches. In warm water, the eggs of most species hatch in 2 or 3 days. Some eggs require a drying period, remaining dormant for months, they hatch soon after water contacts them.

The larvae or "wigglers" feed on tiny bits of organic matter in the water. Many species breathe air through an elongated air tube which they extend through the water surface. Larvae change into comma-shaped pupae, often called tumblers, in about a week.

The pupae transform into adults in about two days. Male mosquitoes feed on nectar of flowers and do not bite. Female mosquitoes also feed on nectar; however, a blood meal is usually necessary to mature the eggs.

38

Mosquitoes show considerable variation in their preferred hosts. Some species feed on cattle, horses, or other domestic animals while others prefer man. A few species feed only on cold-blooded animals and some live entirely on nectar or plant juices. Some are active at night and others only during the daytime.

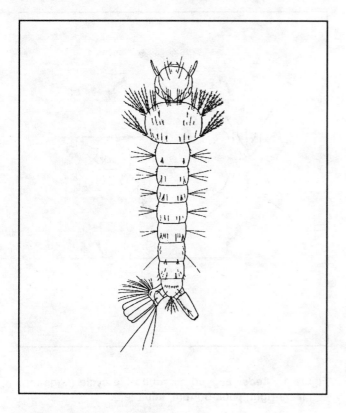

Figure 2. *Aedes aegypti* mosquito larva.

Control

Mosquito control can be divided into two areas of responsibility: individual and public. Individuals are responsible to find and eliminate breeding places on their premises. Receptacles such as old tires, junk automobiles, tin cans, rain barrels, and various plants hold enough water to create mosquito breeding. These can be reduced or eliminated by individuals. To keep mosquitoes from being a problem inside the house, screens should be kept tight fitting and in good repair.

Chemical control of mosquitoes around the home may be accomplished with the use of repellents or space sprays. Repellents are substances that make a mosquito avoid biting people. Several repellents are effective against mosquitoes. All insect repellents

must have the active ingredient appear on the label. Check the label before buying.

Repellent is popular for individual protection because it will repel mosquitoes as well as ticks, fleas, chiggers, sand flies, and black flies. It is available in the form of a liquid, aerosol, lotion, and cream. When applied properly to the neck, face, arms, ankles, and other exposed skin surfaces, most repellents will provide protection from mosquito bites for 2 hours to 12 hours. If desired, old clothing may be sprayed with repellent to provide added protection. Care should be taken not to apply any repellent to eyes, lips or other mucous membranes.

Oil of citronella is another type of mosquito repellent for space repelling. Oil of citronella is the active ingredient in many of the candles, torches, or coils which may be burned to produce a smoke which repels mosquitoes. These are useful outdoors only under windless conditions. Their effectiveness is somewhat less than repellents applied to the body or clothing.

Space sprays may be used to kill mosquitoes present at the time of treatment. The major advantage of space treatment is immediate knockdown, quick application, and relatively small amounts of materials required for treatment. Space sprays are most effective indoors. Outdoors the insecticide particles disperse rapidly and may not kill many mosquitoes. The major disadvantage of space spraying is that it will not control insects for long periods of time.

Mosquitoes can be killed inside the house by using a flit gun or a household aerosol space spray containing synergized pyrethrum or synthetic pyrethroids (allethrin, resmethrin, etc.). Only insecticides labeled for flying insect control should be sprayed into the air. Best results are obtained if doors and windows are kept closed during spraying and for 5-10 minutes after spraying. Follow label directions on the container.

Homeowners may use hand-held foggers or fogging attachments on tractors or lawn mowers for temporary relief from flying mosquitoes. Pyrethrins or 5% malathion can be fogged outdoors. Follow instructions on fogging attachments for application procedure.

Most of the mosquitoes that trouble homeowners and visitors cannot be eliminated through individual efforts, but instead, must be controlled through an organized effort. Florida has over 50 organized mosquito control organizations that specialize in area mosquito control. These control measures include permanent and temporary measures. Permanent measures include impounding water and ditching, and draining swampy mosquito breeding areas. Temporary measures include treating breeding areas to kill larvae and space spraying to kill adults.

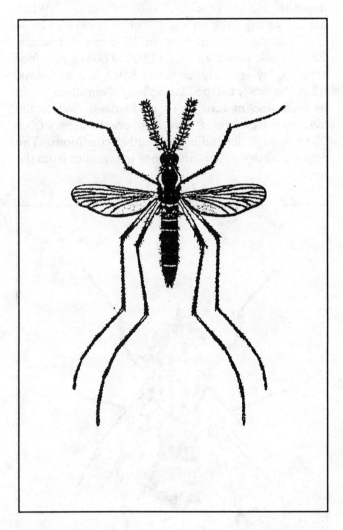

Figure 3. *Aedes aegypti* **adult.**

If you live within an organized mosquito control district, support it in its control efforts. Organized control can accomplish much more than individual efforts. If you are not sure about whether your community has a district, contact the local division of health officials.

SAND FLIES

Often called "punkies," "no-see'ums" or sand flies, the biting midges are vicious. They are often more troublesome than mosquitoes because they can easily enter dwellings through ordinary 16-mesh window screen. The presence of these insects can decrease property values and severely hamper use of recreational areas. Vacationers and campers literally have been driven away from areas by these tiny biters. Sand flies can also cause loss to cattlemen in terms of transmission of various nematode diseases and annoyance to cattle.

Sand flies are members of the insect order Diptera and undergo a complete development with egg, larva, pupa, and adult forms. The adults are less than 1/16 inch long, dark gray to black in color and have one pair of spotted wings. Sand flies breed predominantly in salt marshes; however, some species found inland breed in fresh water areas and tree holes.

Larvae of sand flies are found in mud, sand and debris around the edges of ponds, springs, lakes, creeks and in tree holes or on slime-covered bark. In the water they are free swimmers and are commonly found on floating twigs or leaf trash. The larvae pupate on floating debris or at the water's edge. The adult females, like mosquitoes, require blood to mature the eggs. Males do not bite. Sand fly larvae are in marshes year-round; however, the period of greatest adult activity is June to August.

Sand fly activity is associated with air movement. Adults of most species seldom bite when there is air movement. Sand flies are also sensitive to temperature. Animals having a high body temperature are attractive to great numbers of female sand flies. Persons performing hard labor outdoors frequently are severely annoyed by these insects.

Control

Control of sand flies is too large for individuals to manage. Therefore, mosquito control districts throughout the state are actively engaged in controlling these pests as well as mosquitoes. Excessive problems with biting midges or sand flies should be discussed with the division of health or mosquito control authorities.

40

For temporary relief from annoyance, the homeowner can try several measures. Fogs have been some help when used around the home in shrubbery and other vegetation or other hiding areas. Follow application instructions given on the insecticide label and fogging attachment.

Screening is very important. Since sand flies can get through 16-mesh screen, finer mesh screening should be considered in areas of high infestations. Repellents are suggested for protection from infestations outdoors. When used as directed, a repellent will prevent sand flies from biting for several hours.

BLACK FLIES

Black flies (Simuliidae) are small, dark, stout-bodied flies with a humpbacked appearance. The adult females suck blood mainly during daylight hours and are not host specific. The black fly is a potential disease vector in Florida. It hovers about the eyes, ears, and nostrils of man and animals, often alighting and puncturing the skin with an irritating bite. Black flies are not considered to be major pests of Florida homeowners.

The black fly life cycle begins with eggs deposited on logs, rocks, or solid surfaces in swiftly flowing streams. Larvae attach themselves to rocks or vegetation with a posterior sucker. The length of the larval period is quite variable depending on the species and the larval environment. The adults which emerge after pupation are strong fliers and may fly 7 to 10 miles from their breeding sites.

Control

See mosquito control recommendations.

STABLE FLY

The stable fly, also known as the dog fly or biting house fly, is a blood-sucking fly which closely resembles the house fly. It is similar to the house fly in size and color, but may be recognized by its sharp, piercing mouthparts which project forward from the head. Unlike many flies, both sexes of the stable fly are vicious biters.

The fly is a common pest of man and animals throughout the world. In Florida -- especially western Florida -- stable flies are a serious pest of man and have been a severe threat to the tourist industry.

Stable flies are very persistent when searching for a blood meal and may be easily interrupted in feeding. They may be mechanical vectors of animal diseases but are not considered effective in spreading human disease.

Stable flies breed in soggy hay, grain or feed, piles of moist fermenting weed or grass clippings, seaweed deposits along beaches, and manure. When depositing eggs, the female will often crawl into loose material, placing the eggs in little inner pockets. Each female may lay a total of 500 to 600 eggs in four separate layings. The eggs will hatch in 2 or 5 days, when the newly hatched larvae bury themselves, begin to feed, and mature in 14 to 26 days. While the average life cycle is 28 days, this period will vary from 22 to 58 days, depending on weather conditions. The adults are strong fliers and range many miles from the breeding sites.

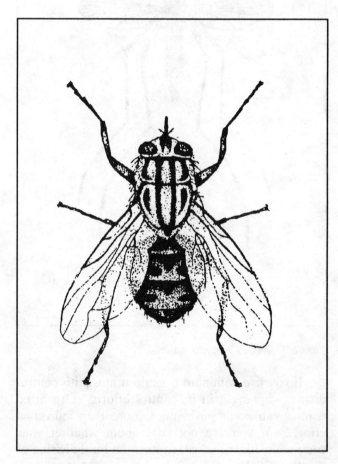

Figure 4. Stable fly.

Control

Stable fly control is most successful with cultural control measures. Since the larvae require moist breeding areas, all breeding sources such as grass clippings should be spread to allow drying.

Homeowners may use fogging attachments on lawn mowers or tractors to provide temporary relief from stable flies. Follow directions on fogging attachments for application procedure.

For personal protection from stable fly bites, repellents may be applied to neck, face, arms, ankles, and other exposed skin surfaces. Most repellents protect for several hours.

If you live within an organized mosquito control district, support its control efforts. Many mosquito control districts have effective programs for stable fly control. Their programs can accomplish much more than individual efforts. If you are not sure if your area is in a district, contact local division of health officials.

HORSE FLIES AND DEER FLIES

Horse flies and deer flies are closely related insects with similar life cycles. Both are strong fliers and only the female bites. They are daytime feeders and can easily cut the skin open for a blood meal.

While feeding the flies inject an anticoagulant into the wound, causing the blood to flow freely. This wound is excellent site for secondary infection. Since these flies are intermittent feeders, they are important transmitters of animal diseases.

Most species of horse and deer flies are aquatic or semi-aquatic in the immature stages. Some will also develop in moist earth, leaf mold or rotting logs. Generally the eggs are deposited in layers on vegetation, objects over water or moist areas favorable for larval development. The eggs hatch in 5 to 7 days and the larvae fall to the water surface or moist areas where they begin to feed on organic matter.

Many species feed on insect larva, crustacea, snails, and earthworms. When the larvae are ready to pupate, they move into drier earth usually an inch or two below the soil surface. The pupal stage lasts 2 to 3 weeks, after which the adults emerge. The life cycle varies considerably within the species, requiring anywhere from 70 days to 2 years.

Control

There is no effective control of the immature stages of horse or deer flies. Individual protection from adults can be obtained by using a standard repellent on exposed skin and clothing before exposure.

Human Malaria

In June 1990, a human case of *Plasmodium vivax* malaria acquired in Florida occurred in a woman camping in the panhandle's Gulf County. This is the first, and only, acquired infection from a mosquito in Florida in 42 years. This factsheet was prepared in response to this case.

Although malaria disappeared as a significant problem in the US by the mid-1950's, it is still one of the most important communicable diseases on a worldwide basis (Figure 1). There were an estimated 489 million cases worldwide in 1986, of which 2.3 million were fatal (Sturchler, Parasitology Today 5:39)

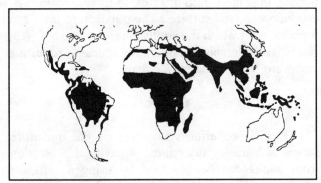

Figure 1. Distribution of malaria.

Malaria in humans is a disease caused by any one of four species of microscopic protozoan parasites in the genus *Plasmodium* (*P. vivax*, *P. falciparum*, *P. malariae* and *P. ovale*) distributed throughout the world.

Overall symptoms may start with headache, aching in the bones, anorexia, and sometimes vomiting. One may feel like the flu is coming on. This is followed by chills, teeth chattering and then sensations of great heat with high fever and sweating, usually in a repeating cycle. If you experience these symptoms and have been in an area were malaria was reported, see your doctor immediately. Malaria can be treated effectively, particularly in the early stages.

Clinically, the malaria infection varies from a moderately severe to a highly fatal illness, depending on the species of parasite, the human's condition, and how soon the patient receives treatment. Malaria caused by *P. falciparum* is particularly severe and often fatal in infants and young children. *P. vivax* generally causes a less severe illness and a lesser rate of mortality. Persons may be infected over and over again, usually developing a tolerance for the parasite which prevents severe illness from reoccurring.

If insufficiently treated, a malaria infection may persist in a person for many months or years and have a continuing or periodically renewed ability to infect mosquitoes, often in the absence of symptoms or with a less severe illness. For *P. vivax* and *P. ovale*, parasites may persist in the liver cells for years and give rise to relapses of the disease by reinvading the red cells during times of stress.

HOW MALARIA IS TRANSMITTED BY MOSQUITOES

The parasite is transmitted from person to person by the bite of *Anopheles* mosquitoes, and ONLY *Anopheles* mosquitoes. The malaria parasite inhabits the human red blood cells, where it multiplies asexually. After reaching maturity in 48-72 hours, the red blood cells burst and release large numbers of new parasites most of which enter new red blood cells; thus, reinitiating the cycle. Others enter liver cells. Before the asexual cycle in the human red cells is established, the parasite must complete at least a 5-10 day period of multiplication in liver cells. The typical malaria symptoms, chills and fever, are associated with this rupturing of infected red cells.

In addition to these asexual forms in humans, some of the parasites develop into sexual forms: the male and female gametocytes. Infection of the mosquito takes place when an *Anopheles* female feeds on an infected person who is carrying gametocytes. The parasite then undergoes a sexual cycle in the mosquito for the next 7-20 days. Numerous microscopic, spindle-shaped forms, known as sporozoites, then invade the mosquito salivary glands. The human infection is initiated when sporozoites are injected during the bite of the infected mosquito.

INTRODUCED MALARIA

Up to now, the only hazard of malaria transmission in Florida stems from people who have relapses, or cases recently acquired in foreign countries where malaria is common. This is referred to as introduced malaria. Despite the widespread presence of *Anopheles* mosquitoes in the US, a highly susceptible human population, and the importation of thousands of cases of malaria acquired overseas, there are relatively few cases reported in the US each year, and very few of those reported were actually infected in the US.

The CDC Annual Summary of Malaria for 1988 (issued in November 1989) reports a total of 1,023 malaria cases in the US. Only 32 of these acquired the infection in this country, none in Florida. Six were fatal. This compares to 932 cases reported in the US in 1987. The geographic distribution of malaria cases in the US during 1988 is shown in Figure 2. In that year, 49 cases of malaria were reported in Florida. However, unlike the case reported in 1990, all the infections were acquired outside the US. They didn't become ill until returning to Florida.

The largest outbreak of introduced malaria since 1952 recently occurred in San Diego County, California. Of the 30 cases between July 24 and September 18, 1988, 28 were in migrant workers and 2 were in local residents who had no apparent malaria risk factors.

FLORIDA'S *ANOPHELES* MOSQUITOES

Of the 70 species of mosquitoes occurring in Florida, 13 are in the genus *Anopheles*. It is easy to recognize adult *Anopheles* mosquitoes by the way they rest on a flat surface, like the skin. Unlike mosquitoes of other genera, *Anopheles* rest with their heads pointed downward and their bodies slanted at a steep angle upward (Figure 3). Other mosquitoes hold their bodies parallel to the resting surface. *Anopheles*

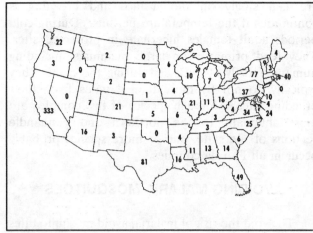

Figure 2. Geographic distribution of malaria cases in the U.S. (1988).

adults also have 3 long mouth parts protruding from the head. Other mosquitoes have 1 long and 2 short mouth parts (Figure 3). The immature stages of *Anopheles* are not easily recognized in the water, as the larvae lie near the surface and are easily confused with floating debris. While all *Anopheles* may be able to transmit malaria, historically, those belonging to the *Anopheles* quadrimaculatus complex of 4 species are considered the important carriers of the disease in the eastern US.

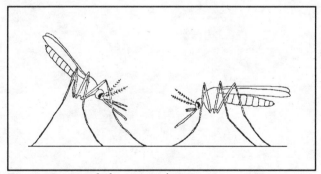

Figure 3. *Anopheles* mosquitoes.

Anopheles quadrimaculatus adults are dark with 4 spots on each wing. They typically breed in permanent bodies of fresh water, e.g. lakes, containing emergent or floating vegetation. The eggs are laid singly on the surface and breeding is continuous if the temperature permits. During cold periods adult females hibernate in protected sites. They feed primarily on large mammals, including humans, mostly at dusk and during the night. They typically do not fly more than 4 miles from their breeding sites. Although species of this complex are most abundant in the northern and panhandle sections of the state, one or more species probably occur in all Florida counties.

AVOIDING MALARIA MOSQUITOES

To avoid the risk of malaria, avoid mosquito bites - it is that simple. Humans cannot get malaria from wild animals, domestic animals or pets. Transmission of malaria from human to human is accomplished by *Anopheles* mosquitoes or by reuse of needles contaminated with the blood of an infected person.

Avoid mosquito bites by staying out of mosquito infested areas, securing window screens, and by applying a repellent containing DEET. Most repellents on today's market contain DEET. DO NOT OVER-APPLY DEET-containing repellent, as this may cause side effects. Some adults have skin reactions to overexposure to DEET, and, in rare cases, children dosed heavily have experienced serious neurological problems, including slurred speech, confusion, seizures and comas. Misapplication of DEET can lead to symptoms similar to malaria and the result could be worse than malaria. PLEASE USE DEET WITH CAUTION.

Eastern Encephalitis - A Fatal Mistake

Eastern encephalitis, also called eastern equine encephalitis and abbreviated EE, is a fatal disease of humans and horses caused by a virus carried by mosquitoes. The disease occurs throughout the eastern United States and Canada from mid-July until first frost in the north, and during most of the year in Florida (Fig. 1). Disease outbreaks are usually limited to 1-3 counties and typically occur every 5-10 years. In some locations, however, there may be horse cases every year. The virus that causes the disease is not normally found in either of these animals, thus the disease is truly an accident of nature.

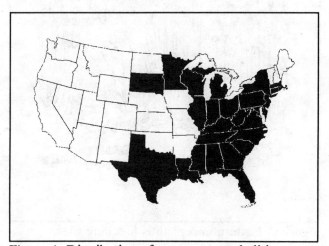

Figure 1. Distribution of eastern encephalitis.

Any Florida horse from Pensacola to Lake Okeechobee that is located near a freshwater swamp or marsh is at greater risk to EE. On the average there are 50-75 horse cases each year in the State. Over 90% of these animals would die if left alone. Mercifully, once a diagnosis is confirmed, most animals are destroyed. There are typically only one or two human cases each year in Florida. Unfortunately, many are fatal; and of those that do survive, nearly all have some residual effects, particularly mental impairment. There is no specific treatment or cure for EE in man or horses but there is a vaccine against the disease in horses.

THE CAUSE

The cause of eastern encephalitis is a virus called eastern equine encephalomyelitis. The virus is normally found only in wild song birds and mosquitoes that live in and around wooded swamps; not just any swamp, but a swamp where there is a certain species of mosquito. The EE mosquito or "black-swamp-mosquito" has the scientific name *Culiseta melanura* (cue-la-see-ta mel-ah-nur-ah).

The virus does not usually cause disease in wild birds, although it may cause a temperature, and perhaps a minor illness similar to a cold or flu in a human. It does not affect the mosquitoes in any way.

MOSQUITO CULPRITS

Culiseta melanura, which translates to "curly black hairs," is indeed a dark mosquito that has a very long proboscis or probe that it uses to draw blood from its hosts. *C. melanura* has very specific breeding requirements. It occurs in most states east and a few states west of the Mississippi River. The larvae are found only in the underwater root systems of deciduous trees that grow in swamps. Fortunately for us, they gets their blood from song birds; rarely does it bite humans or other mammals. And, since *C. melanura* flies no further than 5 miles from its breeding sites, most cases of EE occur within 5 miles of these swamps.

HOW DOES IT HAPPEN?

Well, if *C. melanura* doesn't bite mammals, how do horses and humans get the disease? The entire story is unknown but enough facts have been gathered to construct the following life history:

During warm months when *C. melanura* breeds, there are usually plenty of small birds around for adult mosquitoes to feed on. When female mosquitoes [males do not bite] feed on an infected bird, they pick up the virus. Later, when the mosquito blood feeds on another bird, the virus is transmitted

46

to the new bird. The mosquito remains infected for life and can transmit virus to all birds it feeds on.

After the mosquito blood feeds, the bird becomes infected and the virus begins reproducing in the bird. In a few days, and for only 1 or 2 days, the blood of the bird contains enough virus to infect other mosquitoes that bite it. The bird quickly recovers from the infection and develops immunity. As far as we know, the immunity keeps the birds from becoming infected again. Only newly infected birds can serve as a source of virus for mosquitoes. Therefore, the mosquito seems the most important host as far as virus survival is concerned.

Since *C. melanura* does not bite people, the key to human and horse infection is tied to the short period when birds have high concentrations of virus in their blood. When other mosquitoes feed on infected birds they can become infected as well. It is these "secondary" mosquito species that carry the virus to other vertebrate hosts, including horses and humans. For these secondary mosquitoes to transmit the virus from birds to humans, an individual mosquito must successfully blood feed on both groups of animals. Not all mosquito species do that, *C. melanura* for example.

There is another species of mosquito that is most often associated with outbreaks of EE in horses and humans. This mosquito, the "salt-and-pepper mosquito," has the scientific name *Coquillettidia perturbans* (Coke-qua-la-tid-e-ah purr-tur-bans) or "cokes" for short. This is a large black and white mosquito that looks for blood around dusk. Cokes have a geographic distribution similar to *C. melanura*, but rather than breed in wooded swamps, they breed in cattail or grassy marshes that have a mucky bottom. These types of marshes are often next to the swamps that produce *C. melanura* (Fig. 2).

There may be other mosquitoes, particularly those in the genus *Aedes* (a-e-dees), that also feed both on birds and mammals and thus could possibly transmit EE. The species of *Aedes* involved differ from area to area. Cokes are the only other mosquitoes found throughout the range of EE in the United States. While cokes and *Aedes* can fly more than 5 miles, EE generally does not occur in areas where there are no *C. melanura*.

Cokes and *Aedes* can, and do, take bloodmeals from a variety of other domestic and wild animals;

such as cattle, dogs, cats, squirrels, raccoons, and deer. Fortunately, these animals are resistant to the virus and do not develop EE. Of course, mosquitoes also can blood feed on birds other than wild song birds. Birds that are not native to the U.S., such as ring-necked pheasants and starlings, and some native birds, such as whooping cranes and sandhill cranes can become ill, and many die. Die-off's of exotic birds, particularly pheasants, often precede outbreaks of EE in humans and horses.

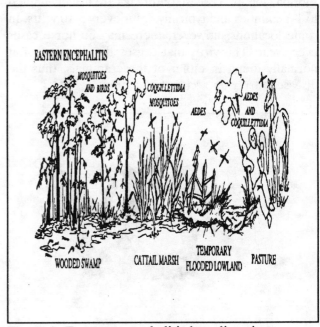

Figure 2. Eastern encephalitis breeding sites.

Unlike wild birds, infected horses and humans normally do not develop high enough concentrations of virus in their blood to infect mosquitoes. This means that they are not good hosts for the survival of the virus. Once the animal recovers from the infection, or dies, the virus in that animal also dies. Animals like this are called "dead-end hosts," not because they die, but because the virus can not be perpetuated. Thus, EE in non-bird hosts is not part of the virus' normal life cycle. It's a fatal mistake for humans, horses and virus alike.

KEEPING A LOOKOUT

Many mosquito control programs in Florida use "sentinel chickens" to alert them when the threat of EE is greatest. Chickens are penned in areas where there are mosquitoes. Every two weeks, a blood sample is taken from some of the birds and tested for antibodies to the virus. Chickens seldom become ill when infected with EE virus. Sentinels were first established throughout Florida in 1978 following a 1977 epidemic of another mosquito- borne viral disease, St. Louis encephalitis (SLE). [There is an IFAS Factsheet on SLE available.] Though chicken flocks were established to monitor SLE, they also have been used to some extent for EE.

While sentinels do not forecast SLE or EE outbreaks, they do provide some limited information on virus activity. Still, the absence of virus activity in chickens does not necessarily mean there is no virus activity. Consequently, sentinel chickens can give a false sense of security regarding EE or SLE activity.

Use of sentinel flocks has been on the decline among mosquito control programs since 1983. Even the Florida Department of Health and Rehabilitative Services, which set up the system, has reduced its support in recent years.

The best system to monitor for EE is to 1) monitor the population levels of the important mosquitoes, C. melanura, cokes, and man- biting Aedes, 2) test these same species for virus, 3) test native song birds for antibodies to EE virus, and 4) establish a system to rapidly detect horse cases near the swamps. Since EE activity in a county during any given year can be limited to a single swamp among many, it is necessary to monitor these factors in all C. melanura breeding swamps.

Florida mosquito control programs typically have a mosquito monitoring system in residential areas. Where there is a potential for EE, the system should extend to the swamps and marshes. Testing mosquitoes and wild birds for virus activity is more expensive than sentinel chickens, but the test results are more reliable for determining the potential for an EE outbreak. Unfortunately, today's testing methods are not readily available or cost effective for mosquito control programs. Frequently, horse cases occur before human cases and are used to indicate the spill-over of virus transmission from the bird-mosquito-bird cycle to the bird-mosquito-mammal pathway.

PREVENTION AND CONTROL

There is a vaccine against EE for horses; and every horse in the state should be vaccinated and have twice yearly boosters. On the other hand, there is no vaccine or cure for EE in humans. The only way to prevent EE in humans is by controlling the mosquitoes that carry the virus. This means controlling cokes, some Aedes, and particularly, C. melanura. Without C. melanura, there can be no EE in horses or humans. It is as simple as that.

It is usually not practical to spray swamps frequently for C. melanura, as is normally done to control other mosquitoes in residential areas. Concerted efforts to reduce mosquito populations in non-residential areas are only made after the State Department of Health and Rehabilitative Services declares an EE alert.

Aborting the threat of an EE outbreak, or an outbreak itself, is most easily done by an aerial application of insecticide to kill adult C. melanura, cokes, and man-biting Aedes. If done properly, only one or two well-timed aerial applications are needed to control the problem for a year. Where aerial application is not possible, adulticides must be applied with the normal ground based equipment used for mosquito control in residential areas.

Aedes can sometimes be controlled in the immature or larval stages, but there is currently no effective way to control larval C. melanura and cokes. Larval control is indicated only for prevention, not as a response to an outbreak.

Ticks

Several species of ticks attack dogs, but cats are rarely infested. Many of the dog ticks are known as wood ticks and infest dogs when they run through the woods or fields. Ticks can also annoy people but humans are not the preferred host.

Ticks are not insects, but are closely related to the spiders. All ticks are parasitic, feeding on the blood of animals. Adult ticks have eight legs.

Of the ticks found in Florida, the brown dog and American dog ticks are the most troublesome. The brown dog tick rarely bites humans, but infestations are frequently found on dogs and in the home. The American dog tick attacks a wide variety of hosts, including humans, but rarely infests homes.

BROWN DOG TICK

The brown dog tick seldom attacks animals other than dogs. It is most likely found where dogs are kept in or around the house. The brown dog tick is not known to transmit diseases to humans but may transmit disease among dogs.

The adult female tick lays a mass of 1000-3000 eggs after engorging on a dog's blood. These eggs are often found in cracks on the roof of kennels or high on the walls or ceilings of buildings. In the house, eggs are laid around baseboards, window and door casings, curtains, furniture and edges of rugs. The egg-laying females are often seen going up walls to lay eggs.

The eggs hatch in 19-60 days into a six-legged, small seed tick. The seed tick takes a blood meal from dogs when they are available. The larvae are so small they won't be noticed on the dog unless a number are together. The seed tick remains attached for 3-6 days, turns bluish in color, and then drops to the floor. After dropping from the host, the seed tick hides for 6-23 days before molting into an eight-legged, reddish-brown nymph. It is now ready for another blood meal and again seeks a dog host.

The nymphs attach to dogs, drop off, and molt to the adult in 12-29 days. As a reddish-brown adult, it again seeks a blood meal, becomes engorged, and is bluish in color, reaching about 1/3 inch in length.

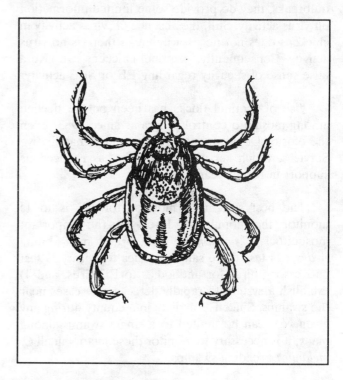

Figure 1. Brown dog tick.

Unengorged larvae, nymphs and adults may live for long periods of time without a blood meal. Adults have been known to live for as long as 200 days without a blood meal. Indoors, ticks hiding between blood meals may be found behind baseboards, window casings, window curtains, bookcases, inside upholstered furniture, and under edges of rugs. Outdoors, ticks hide near foundations of buildings, in crevices of siding, or beneath porches.

AMERICAN DOG TICK

The American dog tick is also a common pest of pets and humans in Florida. The adult males and females are frequently encountered by sportsmen and people who work outdoors. Dogs are the preferred host, although the American dog tick will feed on other warmblooded animals. The nymphal stages of the American dog tick usually only attack rodents. For this reason the American dog tick is not considered a household pest.

The American dog tick requires from 3 months to 3 years to complete a life cycle. It is typically an outdoor tick and is dependent on climatic and environmental conditions for its eggs to hatch.

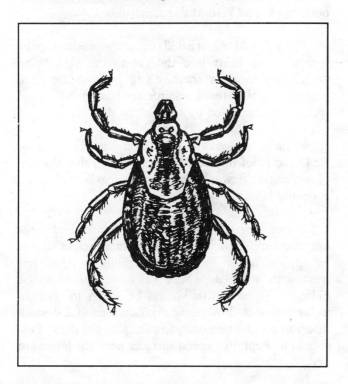

Figure 2. American dog tick.

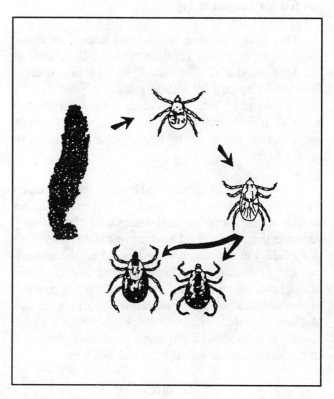

Figure 3. American dog tick life cycle.

The female dog tick lays 4000-6500 eggs and then dies. The eggs hatch into seed ticks in 36-57 days. The unfed larvae crawl in search of a host and can live 540 days without food. When they find a small rodent, the larvae attach and feed for approximately 5 days. The larvae then drop off the host and molt to the nymphal stage. The nymphs crawl in search of a rodent host, attach to a suitable host, and engorge with blood in 3-11 days. Nymphs can live without food for up to 584 days.

Adults crawl in search of dogs or large animals for a blood meal. Adults can live for up to 2 years without food. American dog tick adults and many other species can be found along roads, paths, and trails, on grass, and on other low vegetation in a "waiting position." As an animal passes by, the tick will grasp it firmly and soon start feeding. The males remain on the host for an indefinite period of time alternately feeding and mating. The females feed, mate, become engorged, and then drop off to lay their eggs.

Importance of Ticks

When feeding, ticks make a small hole in the skin, attach themselves with a modification of one of the mouthparts which has teeth that curve backwards, and insert barbed piercing mouthparts to remove blood.

The presence of ticks is annoying to dogs and humans. Heavy continuous infestations on dogs cause irritation and loss of vitality. Pulling ticks off the host may leave a running wound which may become infected because of the ticks type of attachment.

The brown dog tick is not a vector of human disease, but is capable of transmitting canine piroplasmosis among dogs.

The American dog tick may carry Rocky Mountain spotted fever, tularemia and other diseases from animals to people. Dogs are not affected by these diseases, but people have become infected by picking ticks from dogs. People living in areas where these wood ticks occur should inspect themselves several times a day. Early removal is important since disease organisms are not transferred until the tick has fed for several hours.

The American dog tick is also known to cause paralysis in dogs and children where ticks attach at the base of the skull or along the spinal column. Paralysis is caused by a toxic secretion produced by the feeding tick. When the tick is removed, recovery is rapid, usually within 8 hours. Sensitized animals may become paralyzed by tick attachment anywhere on the body.

Lyme disease is transmitted by ticks, but few cases have been reported in Florida. Most transmission occurs in the New England states and the primary vector is the deer tick. The deer tick is not prevalent in Florida, but species that are close relatives and are capable of transmitting Lyme disease are common throughout the state. The American dog tick and the brown dog tick are not considered important vectors of Lyme disease. In cases of tick bites where Lyme disease is suspected, a physician should be contacted so that appropriate blood tests can be done.

Control

Ticks should be removed carefully and slowly from pets and humans as soon as they are noticed. If the attached tick is broken, the mouthparts left in the skin may transmit disease or cause secondary infection. Ticks may easily be removed by touching the tick with a hot needle or alcohol to relax it. Then grasp the tick firmly with tweezers or fingers near the mouthparts and pull evenly and firmly. A small amount of flesh should be seen attached to the mouthparts after the tick is removed.

Pesticidal control of ticks may require treatment of both the pet and the infested area. If a heavy tick infestation occurs it is necessary to treat pets, home and yard at the same time.

Pets should be treated by using dusts, dip or sprays. Rub dusts into the fur to the skin being careful not to allow chemicals to get into the eyes, nose or mouth. Heavy tick infestations on the animal should be controlled by spraying or dipping.

Premise sprays are registered for tick control. Read the label thoroughly to be certain that the site of application (lawn, house, crawl space, kennels, etc.) is on the label.

Brown dog tick infestations of homes and yards are frequently difficult to control. Insecticides should be applied inside the house carefully as light spot treatments to areas where ticks are known to be hiding. Special effort should be given in treating areas frequented by pets. Applications at 2-4 week intervals may be necessary to eliminate the ticks. Pets should be kept off treated surfaces until the latter are dry.

People entering tick infested areas should keep clothing buttoned, shirts inside trousers, and trousers inside boots. Do not sit on the ground or on logs in bushy areas. Keep brush cleared or burned along frequently traveled areas. Repellents will protect exposed skin. However, ticks will crawl over treated skin to untreated parts of the body.

Mange

Mange is an unsightly and painful condition caused by burrowing mange mites. Mange is contagious and spread by contact from infested to non-infested animals. Mange can occur in man, dogs, cats, horses, sheep, cattle and other animals.

CANINE MANGE

Sarcoptic mange of dogs is related to the human skin disease called scabies. Dog mange is caused by the canine mange mite, which frequently also attacks man. A closely related mite attacks cats and produces a severe mange in them.

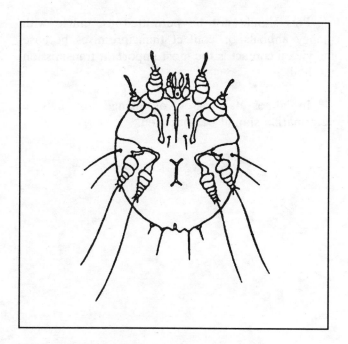

Figure 1. Sarcoptic mange mite.

The female canine mange mite burrows in the upper layers of skin and singularly lays 20 to 40 eggs, which hatch in 3 to 5 days. The larvae molt to the nymphal stage. The nymphs mature to adults. Larvae, nymphs and adult male mites live under scales on the skin surface and do not burrow. The entire life cycle is completed in 8 to 17 days.

Canine mange first appears on dogs as reddish, inflamed papules on the edges of the ears, in the groin or armpits. Usually symptoms first appear on the head. Red spots appear and burrowing female mites cause the skin to exude serum which dries to form scabs and crusts. The scratching of the animals causes the infected area to spread rapidly. Infested areas become dry, hair falls out, the skin thickens and wrinkles. Irritation from the scratching often leads to secondary infection causing an unpleasant odor. If untreated, the animal may die of exhaustion, dehydration or secondary infection.

Canine mange in humans is characterized by a rash that develops after contact with an infected dog. The eruption usually appears as pimples but also may appear as blisters and inflammation. Mange symptoms generally appear on the forearms, thighs and abdomen, but may occur in areas not infested by mites. This is an immune response disease with some individuals reacting more severely than others.

Feline mange usually starts on the heads of cats forming crusts, that cause the skin to thicken and crease.

Dogs and cats exhibiting mange symptoms should be taken to a veterinarian for treatment. Mange symptoms are often confused with flea bite reactions. Humans with canine mange should consult a physician.

RED MANGE

Red mange or demodectic mange of dogs is caused by a mite which lives in the hair follicles of the skin. The first evidence of red mange is the appearance of bald areas where hair has fallen out. As the bald area spreads, itching and irritation increases. Bacterial infections are usually associated with red mange and produce a foul odor. Red mange usually weakens the animal, exposing it to other diseases which then kill the animal. Many animals will self cure. The disease is most common in dogs from 3 months to 1 1/2 years old. Stressed animals often exhibit mange symptoms. The most effective control is applied by veterinarians.

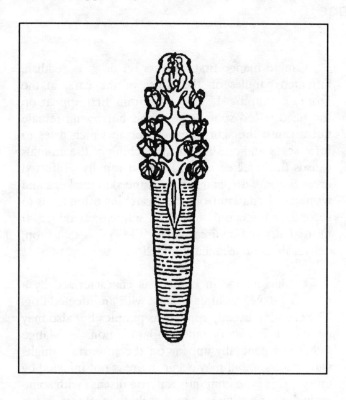

Figure 2. Demodectic mange mite.

EAR MANGE

Ear mange is common among dogs, cats and rabbits. The mites do not burrow in the skin but live deep in the ear canal and feed on skin. The resulting irritation causes the ear canal to become congested. The affected animal rubs its ears and shakes its head to relieve the itching.

Ear mange may be treated by applying mineral oil to the ear canal with a medicine dropper or cotton swab and by cleaning accumulations of foreign matter.

MANGE PREVENTION

Proper care and the maintenance of good health will increase a pet's resistance to skin disease. Canine mange mainly occurs on young animals that are undernourished and suffering from internal parasites and mothered by infested animals.

Pets should not be permitted to mingle with mangy animals or contact their premises because individual contact is the most important transmission method.

In almost all cases of mange on pets a veterinarian should be consulted.

Lyme Disease

Lyme disease was first documented in the U.S. in 1975. The organism that causes Lyme disease is transmitted by tick bites. The disease is named for the town in Connecticut where some of the first cases in the U.S. occurred. Lyme disease cases have since been documented in about 43 states, with over 2,000 cases annually. Several cases of Lyme disease have been reported in Florida.

The primary vectors of Lyme disease in the U.S. are thought to be the northern deer tick (*Ixodes dammini*) in the North and the black-legged deer tick (*Ixodes scapularis*) in the South. Other ticks are also known to transmit the disease. Lyme disease is maintained in wild rodent populations on which the immature ticks develop. These immature ticks pick up the disease organism when they suck the blood of infected rodents. The adult ticks then look for a larger host, such as deer or humans, to get their final blood meal and transmit the disease when they feed.

SYMPTOMS

The first sign of Lyme disease in 60-80% of patients is a red circular rash around the puncture mark where the tick pierced the skin. The most common shape of the rash is an oval 2-3 inches in diameter that usually lasts about 4 weeks. The rash does not itch but may feel warm to the touch. Flu-like symptoms may also develop that often include aches, fever, fatigue, muscle pain, joint pain and headache. Arthritis, cardiac disease and neurologic disorders may develop if the disease is not properly or promptly treated. Sometimes these more serious symptoms develop without the individual ever having a rash.

TREATMENT

Once diagnosed, Lyme disease can be treated. Physicians can determine if an individual has been infected by the Lyme disease organism using a simple blood test; however, some people test negative but have the disease. Infection can be treated by certain antibiotics. However, no immunity is conferred from infection so a person could get Lyme disease again from another bite of an infected tick.

Pets and other animals can contract Lyme disease as well, exhibiting symptoms similar to those in humans. Veterinarians can test for Lyme disease in pets and domestic animals exhibiting suspicious signs of arthritis (in younger animals), heart problems or neurological signs.

PREVENTION

The best prevention against Lyme disease is to avoid being bitten by ticks. Individuals who spend a lot of time outdoors should be aware of the danger and make it a habit to regularly check their bodies for ticks. The tick requires time to attach itself and begin feeding. It is possible to remove ticks before they are able to transmit the disease. Ticks should be grasped with tweezers at the point where their mouthparts enter the skin and pulled straight out with firm pressure. Immature ticks are small and difficult to detect; often they appear as a freckle or mole.

1. Stay out of dense undergrowth unless absolutely necessary. Walk on closely mowed grass or paved walkways whenever possible.

2. Wear long-sleeved shirts and long pants tucked into socks. Light-colored clothes make ticks more visible and easier to pick off.

3. Apply tick repellent to socks and shoes to prevent ticks from crawling onto clothing.

4. Inspect yourself and others thoroughly after walking through areas suspected of being infested with ticks.

5. When taking children on nature outings, keep them in a group with a leader who knows to avoid tick infested areas and can inspect the youngsters for ticks.

6. See a doctor if Lyme disease symptoms appear.

CONTROL

Recommendations for control of ticks include:

1. Keep grass cut low to prevent ticks from developing.

2. Discourage wild animals (raccoons, skunks, deer, mice, etc.) from coming around your yard. They often harbor the ticks that transmit Lyme disease.

3. Although area treatment with insecticides is not warranted in most cases, in some situations it might be appropriate. Check with your county extension office for current IFAS recommendations.

Stinging Or Venomous Insects And Related Pests

Millions of people in the United States are injured from venoms produced by insects and other arthropods each year. Of these injuries about 25,000 result in severe injuries and about 32 result in deaths (Table 1).

VENOMS

Venomous insects and other arthropods produce venoms that can be classified as:

1. Venoms that produce blisters (vesicating toxins) (e.g., blister beetles, certain stinging caterpillars, millipedes).

2. Venoms that attack the central nervous system (neurotoxins) (e.g., black widow spiders, bark scorpions, certain ticks, Hymenoptera, wheel bugs).

3. Venoms that destroy tissue (cytolytic and haemolytic) (e.g., Hymenoptera, fire ants, ground scorpions, mites, chiggers, wheel bugs, brown recluse spider).

4. Toxins that prevent blood from clotting (haemorrhagic) (e.g., lice, fleas, ticks, mites, true bugs, biting flies).

ALLERGIC REACTIONS

Humans differ greatly in their reaction to arthropod venoms. Allergic reactions are often more important than the toxic effects of arthropod venoms. Of individuals who die from arthropod venoms, 96% had an allergic reaction. Allergic reactions to stings or bites are when something happens to a part of the body other than the immediately affected area. Allergic reactions are classified according to the severity:

1. Slight general reaction—inflammation, welts, itching, malaise and anxiety.

2. General reaction—a slight general reaction plus 2 or more of the following symptoms:

swelling, wheezing, abdominal pain, nausea or vomiting.

3. Severe general reaction—any of the above plus 2 or more of the following: difficulty in breathing, difficulty in swallowing, hoarseness, confusion or feeling of impending disaster.

4. Shock reaction—any of the above plus 2 or more of the following: cyanosis, fall in blood pressure, collapse, unconsciousness or incontinence.

Insect stings result in rapid progression of toxic effects. Of 208 deaths, 80% occurred less than one hour after being stung by Hymenoptera (bees, wasps, yellow jackets, or hornets). Spider bites, however, have a longer time interval between bites and death with 89% of 54 victims dying more than 12 hours after being bitten. Statistics also reveal that of snakebite victim deaths, 17% died 1 to 12 hours after being bitten and 64% between 12 hours and 2 days.

What to Do

Insect stings require quick, prompt action.

1. The causative organism must be captured, saved and identified.

2. If a general allergic reaction is suspected, or the victim has a history of hayfever, allergy or asthma, a doctor should be contacted immediately. Additional medical information on poisonings can be obtained from the Centers for Disease Control, Atlanta, Georgia (404-633-3311) or the local Poison Control Center.

3. If marked swelling or discoloration occurs at site of bite or sting, the venom is probably haemolytic, haemorrhagic or vesicating. Keep victim warm and quiet until a physician is reached.

Table 1. Deaths from Venomous Pests (from Parrish AJMS, 1963)

Pests	Number of Deaths Per Year	Percent of Total
Snakes	13.8	30
Insects and Related Pests	32.2	66
Bees	12.4	27
Wasps	6.9	15
Spiders	6.5	14.1
Yellow Jackets	2.2	4.8
Hornets	1.0	2.2
Scorpions	1.0	2.2
Ants	0.4	0.8
Others	1.8	4

4. If little or no swelling or discoloration occurs at the site of bite or sting, the venom is probably neurotoxic. Apply ice to the site or immerse the affected part of body in ice water until a physician is reached.

5. Persons who have exhibited a severe allergic reaction in the past to arthropod venoms or have a history of asthma, hayfever or allergies should:

   a) Undergo skin testing to determine hypersensitivity to arthropod venoms.

   b) Carry identification or tags noting hypersensitivity.

   c) Consider desensitization (immunization).

   d) Carry an insect sting kit (available only with a physician's prescription).

PREVENTION OF STINGS

Several procedures can be used to minimize the danger of being stung by venomous arthropods.

1. Avoid mowing lawns or working with flowering ornamentals when bees and wasps are collecting nectar.

2. Don't walk barefooted in the yard.

3. Sweet items like soft drinks, ripened fruits and watermelons attract bees and wasps. Keep these items covered outdoors. Pick fruit as it ripens and dispose of rotten fruits.

4. Stand still if a stinging insect is near you. If it attacks, brush it off (*don't slap*) to prevent a sting.

5. Control stinging arthropods near heavily used areas.

6. If attacked by a swarm of bees, wasps, yellow jackets or hornets, leave the area immediately using arms to protect your face.

SOME COMMON VENOMOUS ARTHROPODS

Pertinent information on common venomous arthropods is in Table 2.

Bees

Bees are often confused with wasps. Although closely related, they differ in many ways. Bees feed pollen and nectar to their young. They are beneficial insects that pollinate fruits, vegetables and many other plants.

The most common bees are the honey bee, bumble bee, and carpenter bee. Bees are not commonly serious problems and usually require no control.

Figure 1. Honey bee.

When stung by a honey bee, scrape the bee's stinger out of the wound immediately. Be careful not to pull it out. If you do, you will force poison into the wound. If the stinger is not removed, the poison gland attached to the stinger will continue to pump poison into the wound for several minutes. Wasps and other bees do not leave a stinger and are capable of stinging many times.

At certain times of the year (spring and early summer), honey bee colonies divide by swarming. Swarms are not usually a problem unless they land in an inconvenient spot or enter a building. A honey bee colony in a building must be removed after it has been killed to prevent problems from odors of decaying bees, honey and other pests.

If a bee swarm is undesirable in trees, shrubbery or buildings, you may wish to contact a beekeeper, county agent or pest-control company to remove or kill the bees. Insecticide dusts are effective for killing bee colonies in buildings. Dusts may be applied for effective control. To control bees:

1. Locate the colony in the wall at night by tapping and listening for the area of loudest buzzing. Bees keep the nest at 95° so you may be able to feel the heat through the wall.
2. At night, drill a small hole in the wall above the colony and apply dust through it or apply to honey bee entrance to colony.
3. Seal all entrances and exits from the colony.

4. After 2 weeks or when all sound and bee activity has stopped, open the wall and remove dead bees, comb and honey.
5. Bury the colony so valuable honey bee colonies will not be attracted to the residue and be destroyed.

Wasps

Hornets, yellow jackets, *Polistes*, mud daubers and the cicada killers are all wasps. They are generally considered to be beneficial because they attack and destroy many harmful insects found around homes and gardens. Hornets and yellow jackets kill such pests as house flies, blow flies and various caterpillars. *Polistes* are predators of corn earworms, armyworms and many other garden pests. Though beneficial, wasps also attack people. If disturbed, hornets, yellow jackets and *Polistes* will sting. Mud daubers and cicada killers usually are not as aggressive and will not sting unless touched or accidentally caught in clothing. If wasps build nests on houses or in bushes where children play or living activities are carried on, nest destruction or chemical control is necessary.

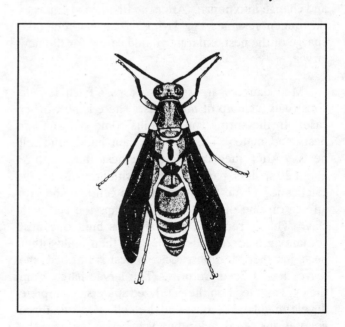

Figure 2. *Polistes* wasp.

Wasps can usually be identified by their nests and their location. Hornets, *Polistes* and mud daubers build nests above the ground. Hornets and *Polistes* nest in trees, shrubbery and under eaves. Mud daubers nest under eaves, porch roofs or similar sheltered areas. Yellow jackets usually build their nests in the ground, but sometimes build them above the ground. Cicada killers nest in the ground.

Hornets and yellow jackets build football-shaped paper-like nests. *Polistes* build paper-like nests that resemble a honeycomb. Mud daubers build clay or mud-cell nests. Cicada killers dig homes about one half inch across and pile the excavated soil around the opening.

Hornets, yellow jackets and *Polistes* are social insects and their colonies develop in a similar way. Adult females make up two castes—queen or fertile females which lay eggs; workers or sterile females which feed larva and may lay eggs without mating if the queen dies during the season. In the fall, queens and males leave the nest and mate. The male dies and the surviving queens hibernate in cracks of rocks, under bark of trees, in buildings, or in the ground. In the spring the queen comes out of hibernation and builds a nest with a few shallow cells. An egg is laid in each cell and these hatch into worker larvae in 2 to 3 days. The queen feeds these larvae which develop in 12 to 18 days and spin cocoon caps over the cells and change into pupae. After the first brood emerges the queen resumes egg laying. The workers take charge of the nest, enlarging it and caring for the new larvae.

Mud daubers are solitary wasps. Each female constructs a clump of mud cells. There is no worker case. In the spring young adults come out of their nests and mate. The females then build mud-cell nests. After they complete the nest they capture about 20 spiders, paralyzing each with her sting as she catches it. These are stored in the cell and she lays an egg on one of the spiders and caps the cell with clay. This is repeated until she has built one nest containing 6 to 20 cells. She may then build other nests in other locations. Once a nest is finished, she leaves it and never returns. The larvae hatch from the eggs and feed on the paralyzed spiders. Complete development takes place in the cell. One to three generations can develop in a year.

The cicada killer is also a solitary wasp. Its habits are similar to the mud dauber except it constructs its cells in the soil and provisions the cells with cicadas.

When a wasp stings it injects a venomous fluid under the skin. The venom causes a painful swelling that may last several days. In some cases a wasp sting may cause severe illness or even death.

Control

Wasps can be easily controlled by applying insecticides to the nest. However, there is usually a certain amount of risk. Nests should be treated at night to minimize the danger of being stung or protective clothing should be worn. Generally, sprays are more effective for aerial nests and dusts are more effective for below ground nests. Residual sprays can be applied for aerial wasp control. Whatever spray is used, it should have a quick knockdown agent such as synergized pyrethrum or pyrethroids mixed with it. Dusts can be applied for control of below ground wasp nests.

Control Procedures

For below ground nests, locate nest and mark area so it is easy to find after dark. Use a flashlight covered with a red cellophane paper so wasps stay in their nest. At night, puff dusts into nest entrance and immediately throw a shovelful of moist soil over entrance. *Be careful not to step into the nest.*

For aerial nests, spray nests with pressurized containers with a pin-stream spray from a distance (20 ft.).

Indoor wasp nests should be controlled as honey bee colonies.

Scorpions

Scorpions are flattened, crab-like animals having ten legs and a flesh tail, ending in an enlarged upturned tip which bears a stinger. They vary in size from one to four inches long. They normally live outdoors, though they will invade homes and buildings.

Scorpions will sting, but usually only when provoked or disturbed. Scorpion venom is a neurotoxin, but the dose injected usually is insufficient to prove fatal to an adult human. None of the several species of scorpions which occur in Florida is capable of inflicting a lethal sting; however, the site of the sting may be sore and swollen for some time.

Scorpions are most active at night. They hide under boards, rubbish, or similar debris which provide shelter and protection. Places commonly infested in a home are under the house or in the attic. They feed on insects, spiders, or similar small animal life.

Figure 3. Scorpion.

Scorpions have a long life cycle. Three to five years may be normal. Males and females go through a courtship ritual prior to mating. Scorpions do not lay eggs and the young are born alive. After birth the young scorpions climb on the back of the mother and remain there until after their first molt. Scorpions are cannibalistic and will readily eat their own species. Females will often eat their own young.

Control

Mechanically destroy any scorpions found indoors by swatting or crushing. Clean out all possible hiding places. Treat hiding or breeding areas with sprays.

Ducks and chickens will eliminate most scorpions from around a building. During dry weather scorpions can be attracted and trapped by spreading moist burlap on the ground around infested areas.

Spiders

Almost all spiders in Florida are harmless to man. Most species do not bite unless provoked to attack. The widow spiders, primarily the southern black widow and northern black widow, are the most frequently found venomous spiders. The brown recluse spider is not considered to be established in Florida although physicians have diagnosed its bites on patients.

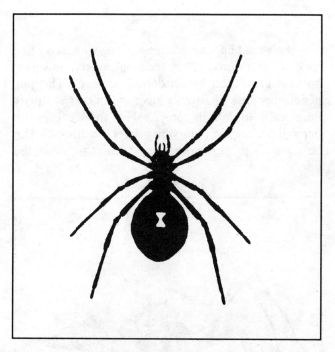

Figure 4. Black widow spider.

Fire Ants

Fire ants look like ordinary house ants; however, they are an aggressive ant capable of inflicting a painful sting. The colony of imported fire ant is a mound sometimes 3 feet across.

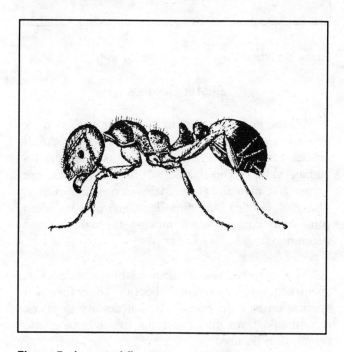

Figure 5. Imported fire ant.

Wheel Bug

The wheel bug is a predacious bug with a cog-like crest on its thorax. They feed on insects; however, humans are bitten by accidental contact. The bug penetrates the skin with its beak and injects a salivary fluid used to kill its prey. The fluid causes an immediate intense pain which lasts 3-6 hours. The best way to prevent wheel bug bites is to avoid the insect.

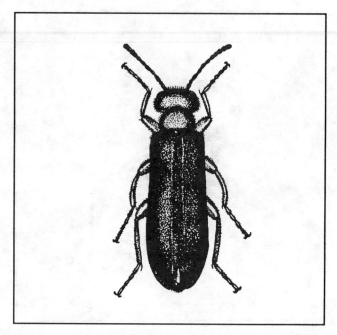

Figure 7. Blister beetle.

Stinging Caterpillars

Stinging caterpillars frequently found in Florida are the puss caterpillar, saddleback caterpillar, Io moth caterpillar, and the hag moth caterpillar. These caterpillars feed on vegetation and have spines which can break off in the skin. When the spines break, a toxin flows from the spines onto the skin, causing a burning sensation.

When working in an infested area, wear protective clothing.

Saddleback Caterpillar

This is a very unusual and striking insect. It is brown with a green back and flanks on which is a conspicuous, brown, oval-shaped central area usually bordered with white. The brown spot gives the appearance of a saddle and the green area appears to be a saddle blanket; hence, the common name. It may exceed an inch in length and is stout bodied. The primary nettling hairs are borne on the back of paired fleshy protuberances toward the front and hind ends of the body. There is also a row of smaller stinging organs on each side. This caterpillar feeds on many plants including hibiscus and palms, but appears to show little host preference.

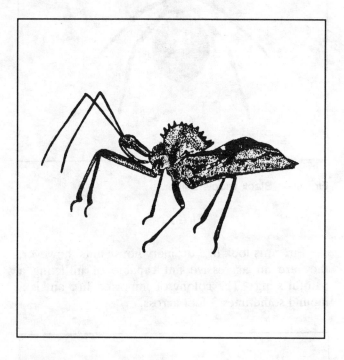

Figure 6. Wheel bug.

Blister Beetles

Blister beetles are narrow beetles with a neck which is slenderer than the head and wings. Adult beetles can release a fluid which causes blisters on human skin. The larvae of blister beetles are harmless to man and are predaceous on other insects. The adult beetles feed on foliage, and persons often come into contact when moving through infested vegetation.

The only suitable control of blister beetles is avoidance of individual beetles or chemical application to crop plants. It is necessary to check recommendations for the crop to determine the chemical to be used.

Puss Caterpillar

It is a convex, stout bodied larva, almost 1" long when mature and completely covered with gray to brown hairs. Under the soft hairs are stiff spines that are attached to poison glands. When touched, these poisonous spines break off in the skin and cause severe pain. Puss caterpillars feed on a variety of broadleaf trees and shrubs, but prefer oaks and citrus. In Florida, there appear to be two generations per year, one in spring and the other in the fall. Natural enemies keep these caterpillars at low numbers during most years; however, they periodically become numerous.

Hag Moth Caterpillar

This caterpillar is light to dark brown in color. It has nine pairs of variable length lateral processes that bear the stinging hairs. These processes are curved and twisted and likened by some to the disheveled hair of a hag, for which it is aptly names. It is found on various forest trees and ornamental shrubs, but is not as common as the other stinging species.

Io Moth Caterpillar

This is a pale green caterpillar with yellow and red stripes. It often exceeds 2" in length and is fairly stout bodied. The nettling organs are borne on fleshy tubercles, and the spines are usually yellow with black tips. They feed on a wide range of plants; however ixora and roses are favorite hosts.

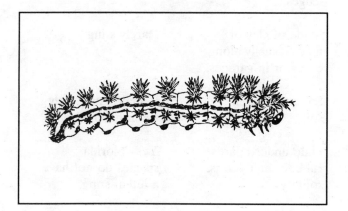

Figure 8. IO moth caterpillar.

Velvet Ant

Velvet ants belong to a large family of wingless, ant-like wasps. The females are solitary, parasitic wasps with an efficient, large stinger. Most species are parasitic on solitary bees and wasp species.

Humans are usually stung by velvet ants when the female wasp is accidentally stepped on with bare feet or trapped against the body in clothing or bedding. Since the wasp is solitary and roaming, control is difficult.

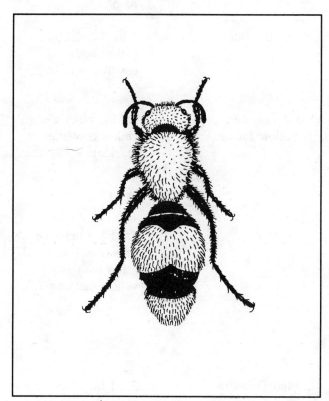

Figure 9. Velvet ant.

Table 2. Summary of Important Stinging Arthropods.

Pest	Description	Nest	Biting or Stinging Ability
Honey Bee	2/3 in., yellow and black or brown, body covered with hairs.	Made of wax cells, found in wall voids or under slab foundations.	Yes
Bumble Bee	1 in., black and yellow body covered with hair.	Made of wax cells, below ground, under slabs, in wall voids.	Yes
Hornets and Yellow Jackets	3/4 in., black with yellow or white markings.	Made of papery material. Nests either aerial or below ground or both. Nests large and globular.	Yes
Polistes	3/4-1 in., black or brown with red and a few with small yellow markings.	Made of papery material. One circular comb of cells opening downward commonly under eaves or window ledges.	Yes
Mud Daubers	3/4-1 in., black and yellow, or metallic blue, or shiny black	Made of clay or mud. Usually along eaves or in garages.	Rarely sting.
Cicada-killer	1 1/2 in., black with yellow markings.	Solitary, digs in soil.	Females sting only when handled.
Scorpion	1-4 in., have 10 legs with tail which bears a stinger.	Hide under boards, rubbish, and debris, solitary.	Yes—Florida species do not have a lethal sting.

Table 2. (cont.)

Pest	Description	Nest	Biting or Stinging Ability
Spiders	Depend on species, 8 legs, 2 body regions.	Widespread and solitary.	Brown recluse and widow spiders are the only dangerous species. Other species are not considered dangerous.
Brown Recluse	Brown with fiddle-like mark on cephalothorax, 6 eyes.		
Black Widow	Black with red hourglass mark on underside of abdomen, 8 eyes.		
Fire Ants	1/16-1/4 in., yellowish to dark red with stinger on end of abdomen.	Mounds 3-36 in. high with surrounding vegetation undisturbed.	Yes
Wheel Bug	1-1 1/2 in., cog-like wheel on top of thorax.	Solitary, occur in vegetation and debris.	Bite only when handled.
Blister Beetle	1/2-3/4 in. beetle with thorax narrower than head or wings.	Solitary	Secrete an irritating substance when disturbed.
Stinging Caterpillars IO Moth, Buck Moth, Puss Moth, Saddleback, Slug Moth,	Depend on species, usually have long and short hairs.	Solitary, occur in vegetation and debris.	Hollow hairs with poison sac. Sharp hairs penetrate skin.
Velvet Ant	1/2-1 in., wingless and ant-like, covered with hair bright red, orange, or yellow.	Solitary on ground.	Sting only when handled or trapped.

64

Removing Honey Bee Nests

Removing honey bee nests from cavities (walls of houses, hollow trees) is a time-consuming, labor-intensive practice that should be undertaken by professionals. Continuous honey bee flight activity to and from a hole in a building is an indication of a nest. Many times, this can be confirmed by listening for bees buzzing inside.

An experienced beekeeper usually can remove bees and combs from easily accessible places like hollow trees, but often bees live in building walls or are tucked away where they are impossible to reach.

Simply killing bees in a cavity with an insecticide can have serious consequences:

- Dead bees and dead brood will decay and produce strong odors.

- Stored honey can absorb moisture and ferment or overheat without adult bees to tend it. This results in burst cappings, producing leaking honey from combs which can penetrate ceilings or walls, causing stains, sticky puddles around doors and windows, and softening of drywall.

The quickest way to remove bees from buildings is to kill them and remove all traces of the nest. In most cases an inner wall or ceiling must be removed, however, calling for the services of a building contractor. It is essential to remove all honeycomb and to plug all holes to be certain there is no way for bees to reenter the area. Any remaining bits of bees wax emit highly attractive odors to swarming bees.

There are a number of ways to kill bees. It is important to exterminate a colony when all bees are on the nest (dusk or dawn). This reduces the number that might be in the field and return to cause problems. Many persons use commercially available wasp and hornet spray for killing the bees. This knocks down the insects quickly and can be used from a distance. Dust formulations of labelled pesticides may also be pumped onto an enclosed nest. There is more and more evidence that soapy water is also a

very good material to use that is inexpensive and relatively environmentally benign. How the bees are killed will depend on the particular situation.

A slower method of honey bee removal which kills fewer of the insects can be used in certain situations. It is based on the principle that bees which leave a building can be prevented from reentering. However, the bees will cluster in a large mass around their previous exit where they are encouraged to enter another colony. Experienced beekeepers do the job best; they are used to bees flying around and to being stung occasionally. The following steps are these steps are recommended:

- From a beekeeper, obtain a one-story hive containing one frame of unsealed brood covered with bees, one frame of honey, and adequate frames of drawn comb or foundation to fill the hive.

1. Fold a piece of window screen to make a cone wide enough at the bottom to completely cover the bees' entrance to the building. This cone is then reduced to about 3/8 inch in diameter. Bend the cone's smaller opening upward.

2. Plug all other holes where bees may enter the building. This is the key to any removal process. All other bee exits\entrances must be sealed!

3. Protect yourself with at least a bee veil and long sleeves (bee gloves are optional) and use a smoker to confuse the bees. Fasten the large end of the screen cone tightly over the entrance.

4. Position the one-story hive as near the cone entrance as possible. It can be positioned on brackets nailed to the building. Place the frames with brood and honey in the center of the hive; place frames of drawn comb or foundation at the sides. The hive entrance should be reduced to about a 1-inch opening

to protect the colony from being robbed by stronger colonies that may be in the area. Bees emerging from the screen cone will not be able to find their way back into the building. Instead, they enter the hive. As bees leave the building and move into the hive, the old colony will grow weak.

5. About 4 weeks later, remove the cone. Bees from the new hive will now be able to enter the building (their previous nest) and transfer the honey to the new hive. The queen in the building is lost along with a few other bees and perhaps some brood. However, with all the honey removed, and so there is little possibility of major odor or honey leakage problems. After the bees have moved completely and the honey has been transferred, close all hoses and cracks to prevent bees from reentering.

The following materials will usually be needed to remove honey bee colonies from buildings.

Bee working supplies
    Veil
    Bee suit
    Gloves
    Hive tool

Queen cage
Smoker
Smoker fuel
Matches
Hive
Hive entrance sealer (1/8" hardware cloth cut to size)
About four empty frames, the rest with foundation
    Cutting/framing tray
    Knife
    Wire and/or string
    Pliers
    Buckets with covers for honey comb and scrap comb
    Burlap bags
    Paint brush
    Dust pan
    Water for drinking and cleanup

Tools for structural work
    Ladder(s)
    Hammer and nails
    Crow bar
    Aluminum and cutters
    Saw (skill + cord, hand, chain + fuel)
    Tin foil for sealing holes
    Scaffold material for hive suspension
    Wire funnel
    Staple gun

Information On The African Honey Bee

1. What is the difference between African, Africanized and European honey bees?

Some 10 races or subspecies of honey bees occur in Africa. One of these, Apis mellifera scutellata, from the central and southern part of the continent, is the predominant parental type introduced into South America. This is the African or Africanized honey bee so often sensationalized in the media. The European honey bee is the insect that commonly occurs in North America. This bee is an amalgam of many races of bees imported over centuries predominantly from Europe. Over the years, a small number of bees from Asia and Africa were also introduced. It was this mostly European bee population that was found in South and Central America, before the purposeful introduction of African bees.

Interbreeding is possible between the African and European races of honey bees, but the amount that has occurred in the Americas is debatable and appears to vary with location. The term "Africanized" is traditionally applied to any crosses resulting from European and African bees in the tropics. However, a more descriptive nomenclature would either be new world African or neotropical African. In temperate areas, the name neotemperate African has been proposed.

2. What is the history of the African honey bee?

Honey bees from Africa were brought to Brazil in the 1950s. It was thought that genetic material from African bees, already adapted to tropical conditions,

could be used to make the European bees already there better honey producers. Unfortunately, some of the introduced bees were released; they were so successful that a large wild population quickly developed. Over the next three decades, this wild population has spread throughout much of The Americas. It has several traits which are of importance to the beekeeper and general public.

3. What is the African bee parental type's most noticeable characteristic?

The African honey bee defends its nest (stings) far more than does the European honey bee. It is thought to be more defensive because it has many biological competitors, including humans, in its native Africa, where only the most defensive honey bees can survive. Apicultural activity in much of Africa is confined largely to bee hunting rather than beekeeping. This has led to limited selection resulting in a population of bees unpredictable in behavior.

By contrast, the European honey bee population in the U.S. has been selected by beekeepers for manageable traits (gentleness, reduced swarming, high honey hoarding). The consequence of this is that the European bee is much more predictable in behavior and defends its nest to a lesser degree than the African honey bee.

4. What other traits can be seen in the African honey bee parental type?

The African honey bee is physically smaller than the European bee and constructs its comb with smaller cells. The African honey bee also swarms (reproduces) and absconds (abdicates its nest) more than the European honey bee. The survival strategy of the African parental type is to put more energy into defensive behavior and the ability to quickly move (abscond) when demanded by either severe predation or harsh environmental conditions or reproduction (swarming). These behaviors make it extremely suited to tropical environments.

The European honey bee's survival strategy is different than that of the African bee; it defends its nest, absconds, and reproduces less in favor of putting energy into producing and storing honey needed to successfully get through long parts of the year (winters) when resources are absent. These behaviors make it suited to temperate environments.

5. What are the traits of the African honey bee in South and Central America?

The African honey bee may show all degrees of behavior expected in either European or African bees. However, conditions in tropical South and Central America have favored the African parental type. Early study suggesting African bees sting to a much greater degree than Europeans has been well documented, but reports that they swarm and abscond far more than Europeans continue to be controversial. Recent research has led to the hypothesis that there has been little interbreeding between the African parental type and the European in most of tropical South and Central America, resulting in a limited hybridized population.

6. What have been the effects of the African honey bee on South and Central American beekeeping?

Bees with predominantly African characteristics have initially brought disaster to much of South and Central American beekeeping. Beekeepers were not prepared for the large, wild population of honey bees which invaded their area; they saw their own bees change rapidly and could not cope with the greater degree of defensiveness. Honey production rapidly declined in areas invaded by wild African honey bees as many beekeepers simply did not adapt and abandoned their apiaries. Those that stayed in business found locations increasingly more difficult to obtain because of the bees' deserved reputation for extreme defensiveness. Finally, there is evidence that the wild African bee population has reduced nectar resources in certain areas, making them unavailable to managed European bees.

Over time, beekeepers have devised ways to manage predominantly African bees, and beekeeping is returning to many areas in South and Central America, although often in a much different form. This is certainly the case in Brazil. Reports from there that African honey bees alone have increased honey production need to be tempered by the fact that beekeeping knowhow has also greatly increased.

7. What has the African honey bee meant to the South and Central American general public?

Like the beekeeper, the general public was unaware of the spread of wild, defensive African honey bees. As the wild population filled a large ecological vacuum and grew in number, it was inevitable that more bee-human contact would result. European bees are fairly selective for nest sites; African bees are not so particular and nested in places where the public did not expect honey bees to become established. Most persons were totally ignorant of bee behavior and those with experience tended to treat all bees like the existant European population.

The defensiveness and erratic behavior of African bees, as well as the fact that many more bees were nesting in a great variety of sites, resulted in a number of stinging incidents involving both humans and animals. These were often sensationalized by the local press. Once the invading wave of African bees passed, gradually the general public and public agencies were able to contend with the fact that a large number of defensive honey bees would be a permanent part of their environment.

8. How will arrival of the African honey bee affect the U.S. beekeeping industry?

Because it will be more informed about the African honey bee, the U.S. beekeeping industry is not expected to suffer extensive damage from the invasion. It is anticipated that beekeepers will attempt to maintain European stock by more frequent requeening and will probably become involved in exterminating wild bee nests to protect their managed bees from resource competition. Therefore, beekeepers will become one of the principle resources the general public and agriculturalists can turn to as they confront the African honey bee.

Sensationalized press releases about any stinging incidents will reduce the number of possible locations for managed bees. This will be alleviated to some extent by beekeepers' effectiveness in convincing the general public that managed colonies will lead to fewer wild, over-defensive colonies. Where wild populations of African honey bees build up, competition for resources with managed colonies will contribute to further reduction in forage availability. All this means more cost to individual beekeeping operations. Thus, the profit margin in beekeeping will become thinner.

As the wild African bee population becomes established, there exists more potential for hybridization in temperate areas. Thus, the characteristics that distinguish the African honey bee from its European cousin in the tropics could become blurred. The extent of hybridization will also effect the northward expansion of the wild bee population. This means that continual rethinking of the bees' effect on humans, animals and the environment will be in order as the invasion spreads into the United States.

9. When is the African honey bee due to arrive in Florida?

It is estimated that the bee will arrive in 1995. To date, African honey bees have migrated about 300 miles per year through the tropics. However, in temperate areas, the rate of spread is less well known. The first detection of the bee was made in Texas in late 1990. Florida is far from that border and the beekeeping industry will profit from knowlege about the bees' spread through intervening states.

10. How will arrival of the African honey bee affect the Florida public?

Experience in Latin America suggests that when the African honey bee arrives, there will be more bee-human contact. Florida's subtropical environment will be favorable to wild populations and the possible effects of publicized stinging incidences could affect agriculture and tourism in the state. Therefore, educating the public and the beekeeping industry will become a priority. This should also lead to a more informed public concerning the value of bees to agriculture.

11. What action should be taken by the beekeeping industry and regulatory officials before arrival of the African honey bee?

Officials and beekeepers should continue to keep themselves aware of the bee's movement in the 1990s. Beekeepers must show the general public that they are the first defense against African honey bees. They can do this by managing their bee stock to maintain gentleness.

Experience in Latin America has shown that wild swarms, though gentle themselves, can develop into large colonies of extremely defensive bees. Thus, the traditional method of hiving or maintaining wild swarms (those from unmanaged apiaries) should be modified. Requeening captured feral swarms with European queens, not yet practiced to a high degree in South and Central America, will quickly be adopted by North American beekeepers. This procedure will

have two benefits; possible elimination of bees that might cause highly publicized stinging incidents and reduction of wild competition for colonies of managed, docile European bees.

The public should be educated to avoid wild bee nests as potentially populated by over-defensive honey bees and to begin reporting wild swarms/nests to police and fire department offices. In addition, information on what provokes defensive behavior and how to minimize the effects of stinging incidences should be disseminated. All established feral bee nests near urban areas should be summarily destroyed by pest control operators, police and/or fire departments and beekeepers.

12. What action should be taken by the general public?

The general public should maintain a healthy respect for all bee colonies and swarms. Any wild swarms found near residences or close to domestic animals (horses, cows, poultry, hogs, dogs) must be suspect and reported to pest control operators, police, fire departments, and/or beekeepers.

This information sheet was developed by the Africanized Honey Bee Task Force made up of personnel from USDA APHIS; Division of Plant Industry, Florida Department of Agriculture and Consumer Services; and The Institute of Food and Agricultural Sciences (IFAS), University of Florida. Thanks to Drs. H. Shimanuki and W.S. Sheppard, Bee Research Laboratory, Beltsville, MD for helpful suggestions. Published March, 1991

List of References:

Anonymous, 1972. Final Report of the Committee on the African Honey Bee. Washington, D.C.: Natl. Res. Counc. Natl. Acad. Sci. 95 pp.

Anonymous, 1986. Proceedings of Africanized Honey Bee Symposium in Atlanta, GA. The American Farm Bureau Research Foundation. Park Ridge, IL.

N.E. Gary, H. Daly, S. Locke and M. Race. 1985. The Africanized Honey Bee: Ahead of Schedule. California Agriculture, November-December, pp 4-7.

McDowell, R. 1984. The Africanized Honey Bee in the United States: What Will Happen to the U.S.

Beekeeping Industry? Agric. Econ. Rep. No. 519. Washington, D.C., U.S. Govt. Printing Office. 33 pp.

Taylor, O. 1985. African Bees: Potential Impact in the United States. Bulletin of Entomological Society of America. Winter, 1985, pp. 15-24.

Spivak, M., D. Fletcher and M. Breed, eds. 1991. The 'African' Honey Bee. San Francisco, CA: Westview Press. 435 pp.

Contact persons:

Dr. Alfred Dietz
APHIS
6505 Belcrest Rd.
Hyattsville, MD 20872
301/436-8716
Lenses for identification;
Model AHB Management
Plan

Dr. H. Shimanuki
Ag. Research Service
BARC East, Bldg 476
Beltsville, MD 20705
301/344-2205
Official AHB
Identification

Dr. Jim Tew
Agricultural Technical Inst.
Wooster, OH 44691
216/264-3911
Federal Leader in AHB
Information;
Fact Sheets & Videos

Dr. Tom Sanford
Bldg 970, Hull Rd.
Gainesville, FL 32611
904/392-1801, ext 143
Florida Action Plan

Laurence Cutts
Division of Plant Industry
P. O. Box 1269
Gainesville, FL 32608
904/372-3505
Florida Bee Regulations

Phyllis Habeck
Florida DPI
P. O. Box 1269
Gainesville, FL 32608
904/372-3505
Public Information

Institute of Food and Agricultural Sciences
Department of Entomology and Nematology

SP 120

Bloodsucking Insects

P. G. Koehler and J. L. Castner[1]

Bloodsucking insects are capable of tormenting humans or animals and transmitting disease. They are all parasites of humans or other host animals and are abundant at certain times of the year. Bloodsucking insects can be grouped as mosquitoes, flies, lice, and true bugs.

Asian tiger mosquito, *Aedes albopictus*. The Asian tiger mosquito is recognized by a straight white stripe in the center of the thorax. It is a daytime biter and frequents shady areas. It breeds in artificial containers and tree holes that retain water. It prefers to rest in low shrubs near the ground.

Black salt marsh mosquito, *Aedes taeniorhynchus*. The salt marsh mosquito is very common in coastal areas and is responsible for most mosquito spraying in Florida. It is not an important disease vector, but emerges in large numbers after rains and flooding.

Stable fly, *Stomoxys calcitrans*. The stable fly looks like the house fly but sucks the blood of man and animals. It has a bayonette-like proboscis that projects forward from the head. The thorax has a light-colored area between the longitudinal bands. The upper abdomen has a number of round, dark spots. It usually bites about 12 to 18 inches from the ground.

Horse fly, *Tabanus spp.* Horse flies are large, heavy-bodied flies with large eyes. The wings are swept back at rest and the abdomen is pointed. The females suck blood and are strong, fast fliers. Larvae live in the mud on the bottom of ditches or in moist soil, feeding on other organisms. At certain times of the year, they are very prevalent.

Deer fly, *Chrysops spp.* Deer flies are often called green-headed flies or yellow flies. Many species are light brown or yellow in color. They suck blood of man and other animals. The larvae live in marshy areas.

Sand fly, *Culicoides spp.* Sand flies are small blood-sucking gnats that are often called noseeums or punkies. Many breed in salt marshes or other moist areas. They are small enough to enter houses through normal window screen. Most species are active at sunrise and sunset.

Head louse, *Pediculus humanus capitis*. Head lice are wingless insects with sucking mouthparts that are 1 to 2mm long. The legs are about the same length with claws to grasp hairs. This louse is found only on the hairs of the head. Eggs are glued to the hairs; preferred sites are on hairs behind the ears and on the back of the head.

Pubic louse, *Pthirus pubis*. The pubic louse or crab louse lives in the pubic region of the body but can spread to hair of chest, armpits, or eyebrows. The crab louse has a wide, oval body and large, grasping claws on the second and third pair of legs. The abdomen is relatively short with hair processes laterally.

Bedbug, *Cimex lectularius*. It is an oval, flattened insect about 4 to 5mm long that hides in cracks and crevices. It has a four-segmented beak that is hidden under the head. Bedbugs are brownish red in color, but may be bright red immediately after feeding. They usually feed at night while the host is sleeping.

Bloodsucking conenose, *Triatoma sanguisuga*. It is a large insect, about 25mm long, with a three-segmented beak. It has a cone-shaped head and a ridge along the edge of the abdomen that has alternating yellow and dark brown areas. The conenose sucks blood at night while the host sleeps. It is often associated with nests of wild animals.

1. Professor and Scientific Photographer, respectively, Department of Entomology and Nematology, Cooperative Extension Service, Institute of Food and Agricultural Sciences (IFAS), University of Florida, Gainesville.

Bloodsucking Insects

Asian tiger mosquito

Black salt marsh mosquito

Stable fly

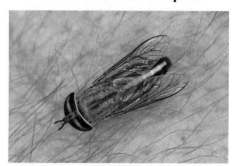

Horse fly

Deer fly

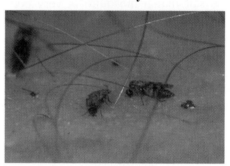

Sand fly

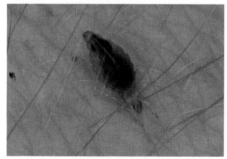

Head louse

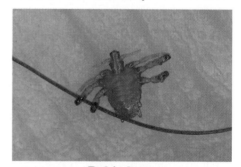

Pubic louse

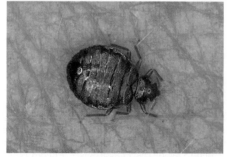

Bedbug

Bloodsucking conenose

Venomous Spiders in Florida

D. E. Short and J. L. Castner[1]

Five species of venomous spiders occur in Florida: The Southern black widow, Northern black widow, red widow, brown widow and brown recluse.

The four species of widow spiders are very similar in body shape. All are about 1 1/2" long with legs extended. Their life cycle is also similar. The female lays approximately 250 eggs in an egg sac which is pear shaped and about 1/2 to 5/8" in diameter. The eggs hatch in approximately 20 days. As the young spiders mature, they construct a loosely woven web and capture progressively larger prey. In Florida, all the widows except the northern black widow breed year-round.

If bitten by a spider, preserve it in rubbing alcohol for positive identification. Most spider bites are not considered dangerous, but if one of the widow or brown recluse spiders are suspected, get medical attention immediately.

Southern black widow. This is the most widespread widow spider in Florida. It is glossy black and has a complete hourglass marking on the underside of the abdomen. The northern black widow has the same general appearance but has two red triangles resembling an hourglass, and a row of red spots on top of the abdomen. The northern species is found west of Tallahassee, primarily in forests with its webs three to 20 feet above the ground. The southern black widow is usually found outdoors in protected places, such as under rocks and boards and in and around old buildings. The bite of the black widow and other widow spiders usually feels like a pin prick. The initial pain disappears rapidly leaving local swelling and two tiny red marks. Muscular cramps in the shoulder, thigh and back usually begin within 15 minutes to three hours. In severe cases, pain spreads to the abdomen, the blood pressure rises, there is nausea, sweating and difficulty in breathing. Death may result, depending on the victim's physical condition, age and location of bite. Death seldom occurs if a physician is consulted and treatment is prompt.

Red widow. This species has a reddish-orange head, thorax and legs with a black abdomen. The top of the abdomen usually has a row of red spots with yellow borders. This spider lacks a complete hourglass on the underside of the abdomen and instead usually has one or two small red marks. The red widow constructs its web in palmettos and has been found primarily in sand-pine scrub habitats in central and southeast Florida.

Brown widow. This spider varies in color from light gray to light brown to black. The abdomen has variable markings of black, white, red and yellow. The underside of the abdomen has an orange or yellow hourglass. It is found most often south of Daytona Beach along the coast. It usually makes its web on buildings in well lighted areas.

Brown recluse. This is not an established species in Florida. It is recognized by the distinctive dark violin-shaped mark located on the head and thorax. The brown recluse is a medium-sized spider about 1/4 to 1/2" in length. It is light tan to deep reddish brown. It is usually found in sheds, garages or areas of homes that are undisturbed and contain a supply of insects to serve as food. Favorite hiding places seem to be in arms and legs of garments left hanging for some time or beds that have been unoccupied for long periods of time. Persons bitten by this spider usually do not feel pain for two to three hours. A blister arises at the site of the bite followed by inflammation. Eventually the tissue is killed, leaving a sunken sore. Healing may take as long as six to eight weeks.

---

1. Professor and Scientific Photographer, respectively, Department of Entomology and Nematology, Cooperative Extension Service, Institute of Food and Agricultural Sciences (IFAS), University of Florida, Gainesville.

Southern black widow

Southern black widow

Red widow

Red widow

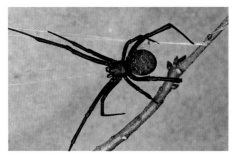

Brown widow

Brown widow

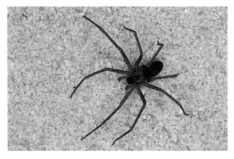

Brown recluse

Brown recluse

Venomous Spiders in Florida

D. E. Short and J. L. Castner[1]

Five species of venomous spiders occur in Florida: The Southern black widow, Northern black widow, red widow, brown widow and brown recluse.

The four species of widow spiders are very similar in body shape. All are about 1 1/2" long with legs extended. Their life cycle is also similar. The female lays approximately 250 eggs in an egg sac which is pear shaped and about 1/2 to 5/8" in diameter. The eggs hatch in approximately 20 days. As the young spiders mature, they construct a loosely woven web and capture progressively larger prey. In Florida, all the widows except the northern black widow breed year-round.

If bitten by a spider, preserve it in rubbing alcohol for positive identification. Most spider bites are not considered dangerous, but if one of the widow or brown recluse spiders are suspected, get medical attention immediately.

Southern black widow. This is the most widespread widow spider in Florida. It is glossy black and has a complete hourglass marking on the underside of the abdomen. The northern black widow has the same general appearance but has two red triangles resembling an hourglass, and a row of red spots on top of the abdomen. The northern species is found west of Tallahassee, primarily in forests with its webs three to 20 feet above the ground. The southern black widow is usually found outdoors in protected places, such as under rocks and boards and in and around old buildings. The bite of the black widow and other widow spiders usually feels like a pin prick. The initial pain disappears rapidly leaving local swelling and two tiny red marks. Muscular cramps in the shoulder, thigh and back usually begin within 15 minutes to three hours. In severe cases, pain spreads to the abdomen, the blood pressure rises, there is nausea, sweating and difficulty in breathing. Death may result, depending on the victim's physical condition, age and location of bite. Death seldom occurs if a physician is consulted and treatment is prompt.

Red widow. This species has a reddish-orange head, thorax and legs with a black abdomen. The top of the abdomen usually has a row of red spots with yellow borders. This spider lacks a complete hourglass on the underside of the abdomen and instead usually has one or two small red marks. The red widow constructs its web in palmettos and has been found primarily in sand-pine scrub habitats in central and southeast Florida.

Brown widow. This spider varies in color from light gray to light brown to black. The abdomen has variable markings of black, white, red and yellow. The underside of the abdomen has an orange or yellow hourglass. It is found most often south of Daytona Beach along the coast. It usually makes its web on buildings in well lighted areas.

Brown recluse. This is not an established species in Florida. It is recognized by the distinctive dark violin-shaped mark located on the head and thorax. The brown recluse is a medium-sized spider about 1/4 to 1/2" in length. It is light tan to deep reddish brown. It is usually found in sheds, garages or areas of homes that are undisturbed and contain a supply of insects to serve as food. Favorite hiding places seem to be in arms and legs of garments left hanging for some time or beds that have been unoccupied for long periods of time. Persons bitten by this spider usually do not feel pain for two to three hours. A blister arises at the site of the bite followed by inflammation. Eventually the tissue is killed, leaving a sunken sore. Healing may take as long as six to eight weeks.

---

1. Professor and Scientific Photographer, respectively, Department of Entomology and Nematology, Cooperative Extension Service, Institute of Food and Agricultural Sciences (IFAS), University of Florida, Gainesville.

Southern black widow

Southern black widow

Red widow

Red widow

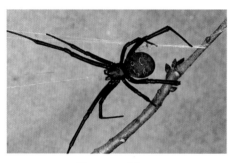

Brown widow

Brown widow

Brown recluse

Brown recluse

Fleas and Ticks[1]

P. G. Koehler, J. F. Butler, and J. L. Castner[2]

Fleas and ticks are the most important external parasites of pets, livestock, and humans. Both fleas and ticks are very abundant, have irritating bites, and are capable of transmitting disease. Fleas are capable of transmitting tapeworms; ticks are capable of transmitting Lyme disease, Rocky Mountain spotted fever, and relapsing fever. Lyme disease is transmitted in the northern U.S. by the deer tick and in the southern U.S. by the black legged, Gulf coast, American dog, lone star, and relapsing fever tick. Lone star and American dog tick can cause tick paralysis.

Cat flea, *Ctenocephalides felis*. The cat flea is the most important flea species in the U.S. and attacks both cats and dogs. Adults are 1/16" long and are usually found on the host. The flea inserts its mouthparts in the skin, injects saliva, and sucks blood. The bite leaves a red spot on the skin. The saliva is irritating to the host, causing dermatitis and hair loss in allergic animals.

Flea eggs and feces. Adult female fleas lay white, shiny eggs on the host. The eggs are not glued to the host so they fall off immediately into bedding and other areas. Adult flea feces are also found in areas the animals frequent. Female fleas can produce 24 eggs per day, and eggs hatch in 12-48 hours.

Flea larva. Larvae of fleas go through 3 stages in 10-14 days; they are about 1/4" long. Larvae prefer to feed on adult flea feces. They are mainly found in pet resting areas, both indoors and outdoors.

Flea cocoons. Flea larvae pupate within silken cocoons and change into the adult stage. Debris is incorporated into the cocoon; therefore they are difficult to find. Fleas can remain in the pupal stage for 6-12 months without a host. They are protected from insecticides and resist chemical treatments.

American dog tick, *Dermacentor variabilis*, (male, left; female, right). This tick is one of the most prevalent tick pests in the eastern U.S. Adults are about 1/4" long, and the shield has variable white mark-ings. The larvae and nymphs prefer to feed on mice, whereas adults prefers dogs and other large animals.

Brown dog tick, *Rhipicephalus sanguineus*, (male, left; female, right). This tick is one of the most common pests of dogs. Adults are 3/16" long and are uniformly reddish-brown in color. All stages prefer to feed on dogs. This tick is prevalent in houses and kennels.

Gulf coast tick, *Amblyomma maculatum*, (male, left; female, right). This tick is very prevalent in the southeastern U.S. The sexes are very different in appearance. The immature stages feed on ground-dwelling birds. Adults primarily attach to the ears of large animals like deer and cattle.

Lone star tick, *Amblyomma americanum*, (male, left; female, right). This tick is one of the most common ticks on humans and has prevented the development of some areas. The female has a silvery spot on dorsal shield. Its long mouthparts allow deep penetration of the skin often causing pus sores.

Black legged tick, *Ixodes scapularis*, (male, left; female, right). This tick is widespread in the southeastern U.S. and often is found along trails, paths, and roadways. Adult ticks are a dark reddish-brown color with dark brown to black legs.

Relapsing fever tick, *Ornithodorus turicata*, (female, left; male, right). The relapsing fever tick is a soft tick and does not have its mouthparts visible from above. The relapsing fever tick mainly attacks rodents, and consequently is associated with rat and mouse habitats. It is capable of transmitting tick-borne relapsing fever.

---

1. This document was published as Fact Sheet ENY-509, Florida Cooperative Extension Service, June, 1991. For more information, contact your county Cooperative Extension Service office.

2. Professor, Professor/Tick Photographer and Flea Photographer, respectively, Department of Entomology and Nematology, Institute of Food and Agricultural Sciences, University of Florida, Gainesville.

Adult cat flea

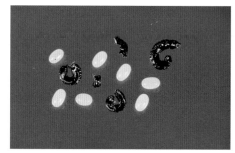

Flea eggs and feces

Flea larva

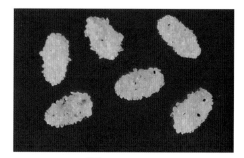

Flea cocoons

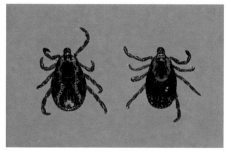

American dog tick

Brown dog tick

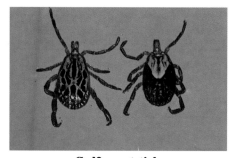

Gulf coast tick

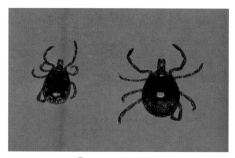

Lone star tick

Black legged tick

Relapsing fever tick

Stinging and Venomous Arthropods Sheet I

D. E. Short, D. H. Habeck, and J. L. Castner[1]

The four major "stinging" caterpillars occurring in Florida are the Puss caterpillar, Saddleback caterpillar, Io moth caterpillar and Hag caterpillar. Less common ones are also discussed. They do not possess stingers, but have spines which are connected to poison glands. Some people experience severe reactions to the poison released by the spines and require medical attention. Others experience little more than an itching or burning sensation.

First aid: Place scotch tape over affected area and strip off repeatedly to remove spines. Apply ice packs to reduce stinging sensation, and followed by a paste of baking soda and water. If an individual has a history of hay fever, asthma or allergy, or if allergic reactions develop, a physician should be contacted immediately.

Saddleback caterpillar. This is a very unusual and striking insect. It is brown with a green back and flanks on which is a conspicuous, brown, oval-shaped central area usually bordered with white. The brown spot gives the appearance of a saddle and the green area appears to be a saddle blanket; hence, the common name. It may exceed an inch in length and is stout bodied. The primary nettling hairs are borne on the back of paired fleshy protuberances toward the front and hind ends of the body. There is also a row of smaller stinging organs on each side. This caterpillar feeds on many plants including hibiscus and palms, but appears to show little host preference.

Puss caterpillar. It is a convex, stout bodied larva, almost 1" long when mature and completely covered with gray to brown hairs. Under the soft hairs are stiff spines that are attached to poison glands. When touched, these poisonous spines break off in the skin and cause severe pain. Puss caterpillars feed on a variety of broadleaf trees and shrubs, but prefer oaks and citrus. In Florida, there appear to be two generations per year, one in spring and the other in the fall. Natural enemies keep these caterpillars at low numbers during most years; however, they periodically become numerous.

Io moth caterpillar. This is a pale green caterpillar with yellow and red stripes. It often exceeds 2" in length and is fairly stout bodied. The nettling organs are borne on fleshy tubercles, and the spines are usually yellow with black tips. They feed on a wide range of plants; however, ixora and roses are favorite hosts.

Hag caterpillar. This caterpillar is light to dark brown in color. It has nine pairs of variable length lateral processes that bear the stinging hairs. These processes are curved and twisted and likened by some to the disheveled hair of a hag, for which it is aptly named. It is found on various forest trees and ornamental shrubs, but is not as common as the other stinging species.

Buck moth caterpillar. This is a large caterpillar, 1 3/4 to 2 1/4" long when mature. It is yellow-brown to purplish-black with numerous small white spots and a reddish head. They feed on oaks, willow, rose and other deciduous plants.

Spiny oak-slug caterpillar. It is a pale green caterpillar, about 3/4" long when mature. Favorite food plants include oak, willow and other deciduous plants.

Flannel moth caterpillar. This caterpillar is about 1" long when mature. Stinging hairs are intermixed with soft hairs in diffuse tufts. Larvae are creamy white, turning dark as they mature. They feed on oak and various other shrubs and trees.

1.   Professors and Scientific Photographer, respectively, Department of Entomology and Nematology, Cooperative Extension Service, Institute of Food and AgriculturalSciences (IFAS), University of Florida, Gainesville.

Saddleback caterpillar

Puss caterpillar

Puss caterpillar

Io moth caterpillar

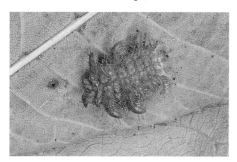

Hag caterpillar

Hag caterpillar

Buck moth caterpillar

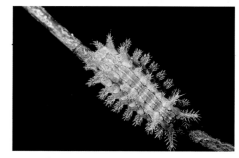

Spiny oak-slug caterpillar

Flannel moth caterpillar

Flannel moth caterpillar

Institute of Food and Agricultural Sciences

Department of Entomology and Nematology

SP 122

Wasps and Bees

P. G. Koehler and J. L. Castner[1]

Wasps and bees can cause problems around structures. Most are social insects that live in colonies. They aggressively defend their nests by stinging. The sting usually involves the injection of a venom that is a nerve poison. The sting may cause death in cases of allergy or when many wasps sting.

Bumble bee, *Bombus spp.* It nests underground in colonies of several hundred individuals. The nests are usually in abandoned rodent burrows, in mulch, or under logs or debris. The abdomen of the bumble bee is covered with hairs.

Cicada killer, *Specius speciosus.* It is 16mm long and black in color with pale yellow markings on the last three abdominal segments. It is a solitary wasp, but colonies of wasps nest in the same location. Each female digs its own hole up to 10 inches deep. It stings and paralyzes cicadas, placing one in the hole with an egg. Closely related species attack and kill flies.

Honey bee, *Apis mellifera.* Honey bee colonies have 20,000 to 80,000 individuals. They are raised for honey and beeswax, and are essential for pollination of crops. The stinger has barbs, so that the stinger and the poison sac remain in the skin. Unlike wasps, honey bees can only sting once.

Honey bee swarm. Most honey bee colonies are in hives that are managed by beekeepers. However, some colonies swarm in large number and can be found outside. They may become established in walls or eaves of houses. The nests in walls can contain a lot of honey. If the bees are controlled or removed, the nest and honey should also be removed to prevent problems to the house.

Mud dauber wasp, *Sphecidae.* It is a black wasp with a long, thin waist, and is not a social wasp. It is not very aggressive and does not sting people often. However, its mud nests are often built close to human activity.

Mud dauber brood chamber. The mud dauber constructs brood chambers from mud on the sides of buildings and under eaves. The wasp stings and paralyzes spiders, lays an egg on them, and seals them inside the chambers. The wasp larva hatches and feeds on the spiders. An emergence hole in the mud means that the wasp has emerged from the chamber.

Paper wasp, *Polistes spp.* It is usually yellow with brown markings or black with red or yellow markings. The wasps are aggressive and readily sting. People are usually stung while trimming shrubbery or cleaning nests from eaves of houses.

Paper wasp nest. Paper wasp nests are made of a papery material that is shaped like an inverted umbrella. It usually has a single comb with up to 250 wasps. Nests are often built under eaves or on branches of shrubs. The eggs are laid in a cell; the larvae hatch and are fed by the wasps. They forage for caterpillars and other small insects to feed the larvae.

Yellowjacket, *Vespula spp.* It is about 12mm long and has alternating yellow and black markings on the abdomen. The wasp is very aggressive in defending itself or the nest. The stinger is not barbed so the wasp can sting repeatedly.

Yellowjacket nest. The nest can be quite large for some colonies of yellow jackets. It is made of a papery material. Inside, the new nest has layers of combs to raise the brood. Some nests are aerial, but most often, the nests are subterranean or both aerial and subterranean. People are usually stung when they step into or disturb a nest.

1. Professor and Scientific Photographer, respectively, Department of Entomology and Nematology, Cooperative Extension Service, Institute of Food and Agricultural Sciences (IFAS), University of Florida, Gainesville.

Wasps and Bees

Bumble bee

Cicada killer

Honey bee

Honey bee swarm

Mud dauber wasp

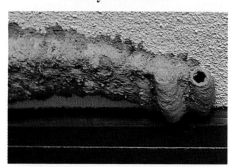

Mud dauber brood chamber

Paper wasp

Paper wasp nest

Yellowjacket

Yellowjacket nest

Insects Around Food

Cockroaches

Introduction

Cockroaches are pests throughout the United States. They are annoying and, when abundant, they are also destructive. Cockroaches, also known as waterbugs, croton bugs or palmetto bugs, destroy food and damage fabrics, book bindings and other materials. When cockroaches run over food they leave filth and may spread disease. They secrete an oily liquid that has an offensive and sickening odor that may ruin food. This odor may also be imparted to dishes that are apparently clean. Excrement in the form of pellets or an ink-like liquid also contributes to this nauseating odor. Some people are allergic to cockroaches and become ill.

Figure 1. American cockroach (actual size, 1 1/2 inches).

Kinds of Cockroaches

The kinds of cockroaches most commonly found in and around Florida homes are Florida Woods Roach, American, Smoky-Brown, Brown, Australian, and German. The smallest cockroaches, the German and the Brown-Banded, are close to the same size and the adults are seldom more than 5/8" long. The larger cockroaches, the American, Australian, Brown, and the Smoky Brown, are 1 1/4" - 2" long and are often called palmetto bugs. Though they are generally found outdoors, they can become an indoor problem when they migrate or are carried indoors. The largest cockroach, the Florida Woods Roach, will also enter dwellings from the outside or from beneath the house. Outdoor cockroaches do not survive well indoors and many times people overreact to the presence of these cockroaches. Often, removal of these outdoor cockroaches from the house is all that is needed for control.

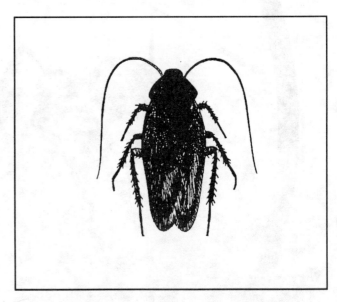

Figure 2. Smoky-brown cockroach (actual size 1 3/4").

Development of the Cockroach

The cockroach has three life stages: the egg, nymph and adult. Cockroach eggs are deposited in groups in a leathery case or capsule called an ootheca. This capsule is usually dropped or glued to some surface by the female as soon as it is formed; however, the female German cockroach carries the capsule protruding from her body until the eggs are ready to hatch. There may be from 30 to 48 eggs in the capsule of the German cockroach, but capsules of other cockroaches may have only 10-28 eggs.

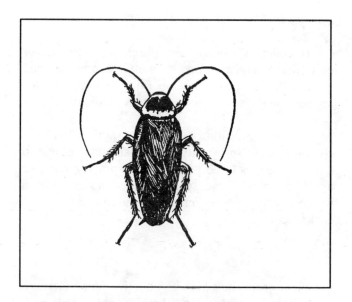

Figure 3. Brown cockroach (actual size 1 1/2").

The newly hatched nymphs have no wings and they shed their skins (molt) several times before becoming winged adults.

German and Brown-banded cockroaches may have several generations per year, but the outdoor species may require a year to develop from egg to adult.

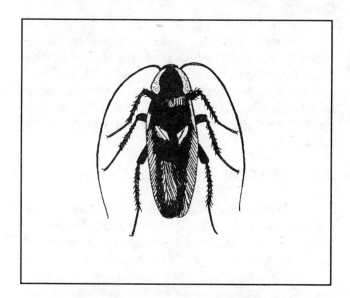

Figure 4. Brown-banded cockroach (actual size 3/4").

Where to Look for Cockroaches

Cockroaches hide in dark, sheltered places during the day and come out to feed at night. They may be found around the kitchen sink or drain board, in cracks around or underneath cupboards and cabinets or inside them (especially in the upper corners), behind drawers, around pipes or conduits (where they pass along the wall or go through it), behind windows or door frames, behind loose baseboards or molding strips, on the underside of tables and chairs, in the bathroom, and in radio and TV cabinets.

The German cockroach is usually found in the kitchen and bathroom, although it may be found all over the house. The other kinds of cockroaches prefer damp, warm places and usually develop in garages, sewers, attics, storerooms and similar locations. They then enter the home from outside breeding sites.

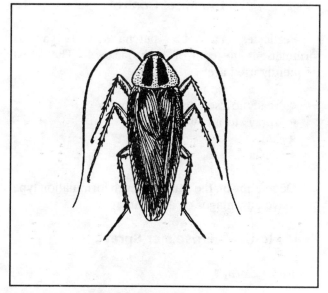

Figure 5. German cockroach (actual size 1 5/8").

Preventive Control

Prevent Infestations

Inspect all baskets, bags or boxes of food, firewood and laundry brought into the house. Destroy any cockroaches or egg capsules. Make it difficult for cockroaches to enter by filling all openings around pipes passing through floors or walls with patching plaster putty, or plastic wood, particularly if cockroaches are coming in from adjoining apartments or from outside. Keep door and window screens in good repair and make sure that there are no cracks between them and the frames. Kill or remove any outdoor species of cockroaches found indoors. Often it is not necessary to spray insecticide to obtain control since outdoor cockroaches don't reproduce rapidly or survive well indoors.

Sanitation

Cleaning will aid considerably in cockroach control. Take away their food supply. Store food in tight containers and avoid spilling flour, cereals and other dry materials in cupboards or on pantry shelves. Do not leave remnants of food on tables or in kitchen sinks overnight. Sweep up any crumbs or bits of food from the floors of kitchen, pantry, and dining area. Put table scraps, vegetable parings and other waste materials in tightly covered garbage cans.

Chemical Control

Pesticides can be purchased in various formulations to control cockroaches. The most frequently used are:

- ready-to-use aerosols or sprays
- sprays to mix with water
- baits
- dusts

Depending on the situation, each formulation type may have advantages or disadvantages.

Ready-to-Use Aerosols or Sprays

Residual Sprays

Ready-to-use sprays are available at local stores as aerosol bombs, bottles with push button sprayers or in containers to be poured into a hand sprayer. These are sold under various trade names with the active ingredient given on the container label.

These ready-to-use products are designed to be sprayed on surfaces to provide residual control of cockroaches. If applied properly, up to 2 or 3 weeks of residual control may be achieved.

Total Release Aerosols

Total release aerosols are useful in reducing large populations of cockroaches but do not give residual action. For best results kitchen drawers and cabinet doors must be open during treatment. Only cockroaches contacted by the spray are killed.

Sprays to Mix with Water

Insecticides may be purchased as either emulsifiable concentrates or wettable powders and mixed with water for spraying. Equipment needed for application is also available at stores selling pesticides.

Water emulsions properly applied with a compressed air sprayer to surfaces should continue to kill cockroaches for 2-3 weeks after spraying. Barrier applications of these sprays around doorways and windows can prevent entry of outdoor species of cockroaches.

Bait

Baits can be highly effective for German cockroach control. The bait is contained in a childproof bait device. Twelve bait devices should be placed in the kitchen, bathroom and other cockroach-infested areas to obtain satisfactory results. The products have no odor and can be applied without removing food, pots, pans or dishes from cupboards. These products are effective for 3-12 months depending on the environment of the home or apartment.

Dusts

The most commonly used dust for cockroach control is boric acid, which is usually applied as 99% active ingredient. It is effective if wall voids and most cockroach harborages are thoroughly treated. Boric acid dust is most effective in dry areas where the dust can be transferred from the treated surface to the cockroach.

Cockroach Control Outdoors

To reduce the number of cockroaches going indoors, it is suggested that baits be applied under the house, porches, etc., and to mulches in and around flower beds, shrubs, etc. Many baits for the control of mole crickets are also labeled for cockroach control. If the house is on a concrete slab, a barrier can be made by applying the bait in a band 1 or 2 feet wide on the ground around the house. Other outside places where cockroaches are commonly found should also be treated. Follow application directions on the manufacturer's label.

Application of Insecticides

Be sure you understand where to look for cockroaches before applying insecticides. Sprays are generally more suitable than dusts and do not leave visible residues. Liquids may be applied with a hand sprayer, push button sprayer, aerosol bomb, or a more expensive compressed air sprayer.

Apply sprays to cracks and crevices, along baseboards, along the back of stove and refrigerator, around pipes, under sinks, around toilet stools, and on exposed surfaces where roaches crawl. Apply sprays on the underside of all objects and any other places where cockroaches may crawl or hide. Apply sprays to the exterior of the house, around windows and doors to prevent entry. Hold sprayer about six inches from the surface being treated and apply a fairly coarse spray so that it does not form droplets and puddles on the floor.

A paint brush is excellent for applying liquid insecticides to baseboards, the insides of cabinets, the bottom and outside of drawers and similar locations. Remove drawers before thoroughly treating the inside of cabinets, desks, and similar furniture. Allow time for cabinet shelves and drawers to dry and then replace shelf paper before replacing contents of shelves and drawers.

A small duster may be purchased for applying dusts to the edges of baseboards, in corners, in and around cupboards and similar hard-to-reach places.

Generally speaking, additional treatments will need to be made in one to two months. Frequency of treatments will depend on sanitation practices, thoroughness of the insecticide application, and how vulnerable the home is to reinfestation.

Insect Growth Regulators

Insect growth regulators (IGRs) are registered for cockroach control. When cockroach nymphs contact surfaces treated with insect growth regulators, they mature into sterile adults with twisted wings. The cockroach population then disappears as the older cockroaches die and no young are born. Control is achieved 6-7 months after the first treatment.

Insect growth regulator applications can be requested of pest control companies and should be applied in conjunction with other registered insecticides. Over-the-counter formulations of insect growth regulators are now available and are mixed with registered insecticides that provide quick kill.

Pest Control Services

The structural pest control industry offers professional services in the control of cockroaches and other household pests. Florida law requires these businesses to be certified and licensed. They have the equipment and training to do a thorough job; therefore, many people prefer using these services rather than trying to control these pests themselves.

Cockroach Control Without Insecticides

Obtaining control of a cockroach infestation is not simply an issue of aesthetics. Cockroaches are capable of carrying disease causing organisms and they are a leading cause of allergies, second only to House Dust mites.

In some instances, persons living in houses treated with insecticide experience reactions to the chemical. The fact that homes remain sealed and the inside environment not subject to normal degradative processes allows pesticides to remain active for longer periods than if they were applied outdoors.

Surveys have shown that more than 75% of urban respondents use 1/3 of all the pesticides applied in the U.S., and most of these pesticides are applied in the home. Over the past few years the general public has become keenly aware of the need for a clean and uncontaminated living environment. There is a need for alternative means of insect control, especially in the home. The following are several key points necessary for the control of cockroaches without the use of insecticides.

THE PRINCIPLE COCKROACH SPECIES

Of the 41 cockroach species in Florida, only about 6 are considered common pests. The predominant pest cockroach species in Florida (and the world) is the German cockroach, *Blattella germanica*. Adults are about 3/4" long and have 2 dark longitudinal bands on their back near the head. Immatures, or nymphs, are smaller, wingless and dark brown with a series of white sections along the margins of the abdominal segments. Adults live about 3-4 months, and during this time females produce about 7 egg cases (oothecae) containing 30-40 nymphs each. German cockroaches are the most prolific of the pest cockroach species.

The remaining 5 species are more closely related to each other than they are the German cockroach. They are the American, *Periplaneta americana*, Australian, *P. australasiae*, Brown, *P. brunnea*, Smokybrown, *P. fuliginosa*, and the Florida Woods cockroaches, *Eurycotis floridana*. Collectively, this group of cockroaches is what the homeowner commonly refers to as "palmettobugs".

There are many differences between this group of related cockroaches and the German cockroach. First, the size, both weight and length, of these cockroaches is many times that of the German cockroach. These 5 larger species do not normally breed indoors, however, in some situations they will establish breeding populations in attics and wall voids. They normally breed outside in sewers, dumps, trees (especially palms), tree holes, mulch, and flower beds, and only enter homes as part of foraging activity for food and especially water. Some of these larger species can live up to 2 years in the adult stage. Females of this group produce an egg case about every 1-2 weeks containing 15-20 nymphs each. A typical female will produce about 20-80 oothecae during her lifetime. In contrast to German cockroaches, these larger species glue their egg case in a hidden, moist, dark area soon after it is developed. German cockroaches carry the ootheca until just before (less than one day) it hatches.

PREVENTING INFESTATIONS

Long term prevention of cockroach infestations without insecticides is the best means of ensuring a cockroach free environment. Not only will the following measures help to prevent a future infestation, they will also help in reducing a present infestation. The following 4 points target the elimination of the most important aspects of cockroach establishment and survival.

Exclusion

(1) One of the primary means of initial cockroach infestation is via roach infested grocery bags. Grocery stores are large, and infestation of products is likely. Make an effort to visibly scan groceries for cockroaches before putting the groceries away.

(2) Guests (adults and children) can often transport cockroaches from their infested home to yours either on themselves or in packages.

(3) Keeping doors and windows shut and screens in good shape can often prevent the entrance of cockroaches into your home.

(4) Caulking cracks and gaps will keep a seal around doors and windows and will help prevent cockroaches from entering your home.

(5) Children can transport cockroaches from school to home in bookbags and lunchpails. Inspect these items on a regular basis.

(6) If a homeowner lives next door (separated by only a wall, such as in an apartment or duplex) to someone with a cockroach infestation, an infestation can be maintained in the clean home from a next door neighbor's "reservoir" of cockroaches. Cockroaches are easily capable of migrating on plumbing from one apartment to the next.

Elimination of Food Sources

German cockroaches can remain alive for approximately 2 weeks with no food or water and for 42 days with only water. It is of utmost importance to realize that cockroaches do not need a large, visible amount of food to survive. The elimination of food sources includes:

(1) Proper storage and prompt disposal of filled garbage cans. Researchers have determined the highest concentration of cockroaches is near the garbage can.

(2) Garbage cans with a sealed lid are helpful in preventing cockroaches access to food sources.

(3) Keeping the garbage can and the area around the can clean are also essential. A frequent wipe-down with a wet rag is often helpful.

(4) Frequent dumping of sink strainers prevents potential food buildup.

(5) Immediate dish washing prevents cockroaches from utilizing crumbs on the dishes. Dishes left unwashed are a major source of food for a kitchen infestation.

(6) Kitchen appliances (toasters, toaster ovens, microwaves, ovens, stoves, and refrigerators) should be kept free of crumbs. In addition, the voids under and behind each appliance should be crumb free.

(7) If pets are present, dry food should be kept in resealable containers. Do not leave food and water out all the time.

(8) Feed your pet at particular times and clean up after every meal.

(9) All foods should be resealed after opening, or kept in the refrigerator.

(10) Periodic sweeping/vacuuming under furniture (mainly dining table) where people may eat (i.e. in the living room in front of T.V.) helps in eliminating food sources.

(11) Periodic shelf emptying and sweeping with a wisp broom not only cleans up spilled food, it disturbs the cockroaches.

Elimination of Water Sources

The single most important factor in determining cockroach survival is availability of water. German cockroaches can remain alive for only 12 days with an abundant food supply and no water, but can remain alive for 42 days with only water. The incidence of large species sightings in homes often increases as periods of drought persist, because these cockroaches are actively searching for sources of moisture. The elimination of water sources includes:

(1) Tightening loose pipes and patching all plumbing leaks both in the kitchen and bathroom.

(2) Indoor plants should not be over-watered, because accumulated water is made available to the insects.

(3) Another common source of moisture is the condensation from under the refrigerator. If these areas can be reached they should be frequently wiped dry or pans placed under the appliance to collect the water. Pans should be frequently emptied.

(4) Pet drink dishes, aquariums and pipe condensation (under sink, in wall voids) can also be sources of moisture.

(5) Moisture sources also are kitchen and bathroom sinks, leaky dishwashers and bathtubs. Steps should be taken to minimize excess moisture buildup in each case, even if this means drying out sinks and bathtubs after use.

(6) Outside, steps should be taken to eliminate places where water collects (tires, cans, tree holes etc...).

Although the elimination of water sources is the simplest goal of the three, in theory, it is often the most difficult to achieve, because leaky pipes in wall voids cannot be detected by the homeowner.

Elimination of Hiding Places (Harborages)

The third critical element for cockroach survival in a home is harborage. By nature, cockroaches run from the light. Cockroaches prefer dark, tight, warm, and moist places. It is no surprise, then, that homes with a cockroach infestation have large numbers under the sink area (dark and close to food and water). The elimination of harborages is important in controlling infestations.

(1) "Cracks and crevices" (the void where 2 separate wall pieces meet) should be sealed with a tube of caulking. Any small gap or hole that leads to a void is a prime cockroach harboring area and must be sealed. Adult cockroaches can fit into cracks only 1.6 mm wide (about 1/16 of an inch).

(2) Uncluttering what is cluttered removes potential harboring areas from cockroaches, so keep all areas of the home, especially the kitchen, uncluttered and free of useless debris.

(3) Outside, remove cluttered debris and trash from around the house.

(4) Stack firewood well away from the house, as this is a prime harborage area for large cockroaches.

(5) Filling in tree holes with cement also removes this prime harborage.

(6) Keep shrubbery and ornamentals well trimmed.

(7) Keep palm trees free of loose and dead palm branches and remove all palm debris.

REDUCING INFESTATIONS
Traps

There are several non-chemical tactics available for reducing an existing cockroach infestation. The first involves the use of traps. Sticky traps (i.e. Roach Motel type) can be purchased and placed, indoors, near the garbage, under the sink, in the cabinets where food is stored, under and behind the refrigerator, and in the bathroom. Outdoors, sticky traps are not recommended because they generally do not hold the larger cockroach species and are not resistant to weathering.

A second use of traps involves baited jars. Any empty jar (pickle, mayonnaise, peanut butter etc...) with a rounded inside lip will suffice. Grease the inside upper lip with a thin film of vaseline (to keep them from escaping) and place a quarter slice of beer-soaked bread into the jar. Any human food will do if you do not have beer. Wrap the outside walls of the jar with a paper towel so the roaches have a surface to grasp as they climb up the outside of the jar. Placing the jar against the wall will also provide easy access of the bait to the cockroaches. To get rid of the live, trapped cockroaches simply pour dishwashing detergent into the jar and add hot water. The cockroaches can then be dumped outside or in the garbage. Wash out the jar and repeat the process every 2-3 days.

Jars should be placed in the same inside areas as the sticky traps. For outdoor trapping, place the jars in trees and tree holes, in the mulch next to the house, and near the garbage disposal and air conditioning pad. Cover the jars with a dome shaped piece of aluminum foil taped to the sides. This will prevent rain from filling the traps.

Biocontrol

Biological control of cockroaches is generally not realized as a source of cockroach control because it is usually present and thus taken for granted. Almost all animals have natural enemies that attack and kill them. Cockroaches are no exception. The conservation of cockroach natural enemies will help keep infestations in check. Natural enemies include wasps (microscopic parasites), nematodes, spiders, toads and frogs, centipedes, birds, lizards, geckos, beetles, mantids, ants and small mammals (mice).

Parasitic wasps are the most important natural enemy listed. There are a number of wasp species ranging in size from less than 1/16 to 1", and they kill cockroaches by drilling a hole into the cockroach egg case and depositing their own eggs, which hatch and eat the developing cockroach embryos.

Cockroach parasitic nematodes are microscopic (<0.5 mm) roundworms of the genera *Steinernema* and *Rhabditida*. Parasitic nematodes occur in very large numbers. Nematodes enter the cockroach gut either through the cockroaches' air tubes or directly through the mouth. Nematodes attach to the cockroach exoskeleton and the cockroaches groom them off. Once inside the cockroach gut the nematodes release a bacteria which multiplies. It is on these bacteria that the nematodes feed and multiply. A single cockroach can supply thousands of nematodes for the next generation. Spiders, toads, frogs, other insects and small mammals feed on a wider array of items, cockroaches of which are only one. There is, however, a large species of spider found throughout Florida that feeds almost exclusively on cockroaches.

One unifying characteristic of all cockroach natural enemies is their susceptibility to toxic pesticides. In most animal species there exists a few individuals which are resistant. Pesticides kill off the susceptible majority and leave only the resistant few. As these resistant few breed the gene for pesticide resistance increases and thus in each subsequent generation a greater majority of individuals are resistant. The phenomena of resistance development can occur only if the population is targeted, as are cockroaches, as a pest. Since beneficial insects are not normally considered pestiferous, they are not the target of insecticide application and do not normally develop resistance. Preservation and perpetuation of the complex of cockroach natural enemies depends heavily on the non-use of pesticides.

Dusts

Indoors, non-organic dusts, such as silica gel and boric acid, can be placed into out-of-the-way "cracks and crevices". The dusts can be applied with a bulbous type duster and can be placed into "cracks and crevices" under the sink area, under the stove and refrigerator, behind baseboards, in electrical outlets and in cabinet cracks. Silica gel is simply finely ground sand or glass that rubs the protective waxes off the cockroach cuticle resulting in cockroach dehydration. Boric acid is also a dust that's placed into "cracks and crevices". Cockroaches walking across an area dusted with boric acid pick up the dust on the sticky cuticle. As the cockroaches groom off the boric acid dust they ingest it into their stomach and it kills them.

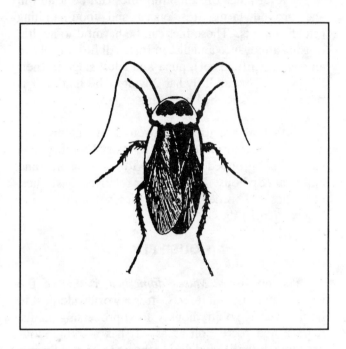

Figure 6. Australian cockroach (actual size 1 1/2").

Ultrasonic Devices

Extensive research has shown that ultrasonic devices neither kill nor repel cockroaches.

In conclusion, reduction of a cockroach infestation without using insecticides can occur only if the insect is first prevented, or excluded, from establishing a population. Once a population is established the elimination of food, water and harborage resources must be performed. This, in many cases, involves a change in lifestyle by the homeowner.

Not only is elimination of food, water and harborage important, but with the use of traps, preservation of natural enemies, and application of non-organic dusts, a reduction in cockroach numbers will be achieved. The integration of these pest management strategies (IPM) will prevent future and reduce current infestations of the 6 principle pest cockroach species in Florida.

Filth-Breeding Flies

Several kinds of non-biting flies can be found in and around farms, residences and food-handling establishments. These flies can be harmful to health, causing annoyance and discomfort. All filth flies have an egg, larva (maggot), pupa and adult stage in their life cycle. The adult fly has 2 wings (the hind pair is reduced to a knobbed balancing organ).

Filth flies are usually scavengers in nature and many can of transmit diseases to man. Filth flies can usually be grouped according to their habits and appearances: houseflies and their relatives; flesh flies, blow flies and bottle flies; filter flies, soldier flies, and vinegar (fruit) flies.

HOUSE FLY

The house fly, *Musca domestica*, is one of the most common of all insects. It is a worldwide pest in homes, barns, poultry houses, food-processing plants, dairies, and recreation areas. It has a tremendous breeding potential and during the warmer months can produce a generation in less than two weeks. In many areas of Florida the house fly breeds throughout the year.

House fly eggs are laid in almost any type of warm organic material. Animal or poultry manure is an excellent breeding medium. Fermenting vegetation such as grass clippings and garbage can also provide a medium for fly breeding. The whitish eggs, which are laid in clusters of 75-100, hatch within 24 hours into tiny larvae or maggots. In 4 to 6 days the larvae migrate to drier portions of the breeding medium and pupate. The pupa stage may vary in length considerably, but in warm weather can be about three days. When the adult emerges from the puparium, the wings are folded in tight pads.

The house fly crawls about rapidly while the wings unfold and the body dries and hardens. Under normal conditions this may take as little as an hour. Mating occurs immediately. A house fly may go through an entire life cycle - egg, larva, pupa to winged adult - in 6 to 10 days under Florida conditions. An adult house fly may live an average of 30 days. During warm weather 2 or more generations may be produced per month. Because of this rapid development rate and the large numbers of eggs produced by the female, large populations build up.

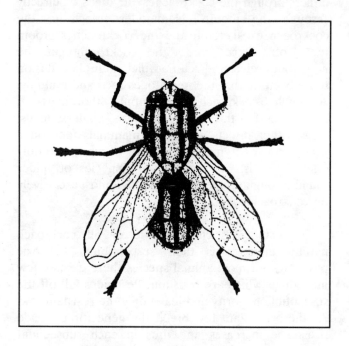

Figure 1. House fly.

House flies are strong fliers and can become widely distributed by flying, wind currents, vehicles and animals. Generally, however, flies are abundant in the immediate vicinity of their breeding site. Under certain conditions, they may migrate 1 to 4 miles, but are usually limited to one-half to 2 miles.

House flies feed by using sponging type mouthparts. As the fly moves about from one food source to another, it samples and eats its food by regurgitating liquid and dropping it on the food to liquify it. Light colored spots called fly specks are visible signs of this type of feeding. Darker fly specks associated with house flies are fecal spots.

The house fly's filthy habits along with its persistence for invading homes and feeding on human food enable the house fly to spread many intestinal diseases such as dysentery and diarrhea.

Resistance is a complex problem associated with chemical control of insects. Recognized some 50 years ago, it is not a new problem. Defined, resistance is the ability of an insect population to withstand exposure to insecticides. This is acquired by breeding from insects that have survived previous exposures to an insecticide that did not wipe out the whole population. The surviving insects breed and develop a resistant strain that survives insecticide treatment.

STABLE FLIES

The stable fly or dog fly is a blood-sucking fly which is of considerable importance to people, pets, agricultural animals, and the tourist industry in Florida. Stable flies primarily attack animals for a blood meal, but in the absence of an animal host will also bite man.

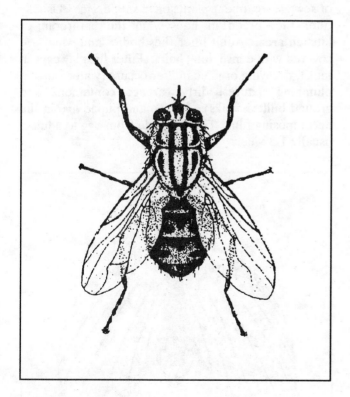

Figure 2. Stable fly.

Adult stable flies can fly up to 70 miles from their breeding sites. The stable fly adult is similar to the house fly in size and color. The stable fly, however, has a long bayonet-like mouthpart for sucking blood. Unlike many other fly species, both male and female stable flies suck blood.

The stable fly breeds all year in Florida although peak populations occur from August through September or October.

Stable fly bites are extremely painful to both man and animal. When hungry, stable flies are quite persistent and will continue to pursue a blood meal even after being swatted at several times. Although the bite is painful, there is little irritation after the bite, and few people have allergic reactions.

The most practical and economical method for reducing stable fly populations is proper management or elimination of breeding sources. It is important to remember that stable flies cannot develop in dry materials.

Stable flies breed in three principal types of material:
1) Green vegetation (e.g. green chop, silage, crop residues, and cut grass)
2) Seaweed and bay grass deposits
3) Animal manure

FLESH FLIES

Flesh flies are a scavenger fly species that usually feeds on carrion or meat scraps in garbage. They are medium-to large-sized flies and usually have 3 dark thoracic stripes and mottled abdomens. Many of the common species have a red tip on the abdomen.

Female flesh flies retain eggs within their bodies until the eggs are ready to hatch. The larvae are deposited directly onto the food the immature will be eating. The life cycle for the common species can be completed in 8 to 21 days.

The preferred breeding media around residences are decayed flesh, spoiling meat, and manure. Usually garbage can meat scraps and dog food left outside are abundant sources of flesh fly breeding. Flesh flies can also breed in dead rodents and birds in attics or wall voids.

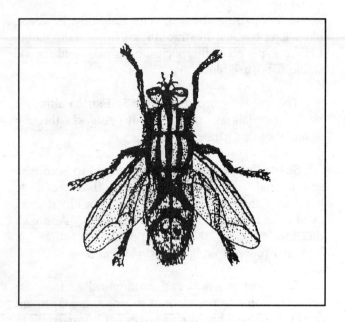

Figure 3. Flesh fly.

BLOW FLIES AND BOTTLE FLIES

There are quite a few species of blow flies and bottle flies found in and around residences. Greenbottle, bluebottle, and bronzebottle flies are particularly abundant in Florida. In urban areas these flies may be more abundant than house flies.

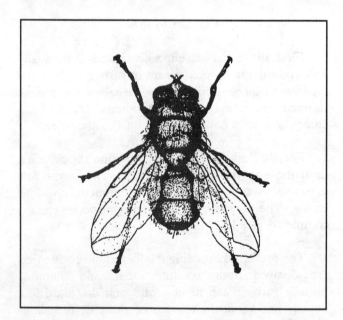

Figure 4. Bottle fly.

Blow flies and bottle flies usually have a metallic blue or green color or both on the thorax and abdomen. These flies are strong fliers and range many miles from breeding places. They are abundant during the warm summer months.

Blow flies and bottle flies can breed on dead rodents and birds in house attics or wall voids. They usually breed in meat scraps, animal excrement, and decaying animal matter around houses. The adult flies are quite active inside and are strongly attracted to light. The mature larvae migrate from breeding areas to pupate and can become a problem when they crawl into a house.

Blow flies usually lay eggs on dead animals or decaying meat. Garbage cans have been known to produce 30,000 blow flies in one week. The life cycle usually lasts 9-21 days from egg to adult.

FILTER FLIES

Filter flies, (drain flies) belong to the moth fly family. They usually feed on slime in trickling filters of sewage treatment plants or in sink drains. Usually filter flies breed in houses, in the bathroom or kitchen area. Adult filter flies bodies and wings are covered with dense, long hairs. Filter flies lay eggs in moist, decaying organic matter occurs (water traps in plumbing fixtures, dirty garbage containers, and around built-in sinks). Larvae and pupae live in the decomposing film. The life cycle from egg to adult is usually 1-3 weeks.

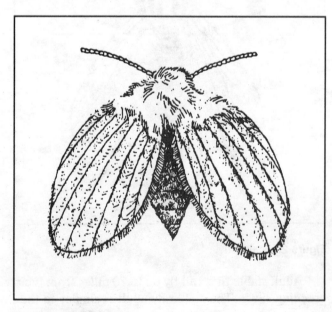

Figure 5. Filter fly.

Filter flies can be controlled by locating and correcting the source of moisture. Timed aerosol space treatments with pyrethrins can break their life cycle.

SOLDIER FLIES

The soldier fly is a widespread pest that occasionally becomes a problem in homes. The adult is a large fly about 1 inch long with 2 large translucent areas on the abdomen. The fly behaves like a wasp and is similar in appearance to a mud-daubber wasp. The larvae prefer to feed on human or animal excrement although they have also been known to breed in honey bee colonies killed in walls of houses.

Most frequently the larvae are found in bathrooms, migrating from the septic tank or sewer line. The presence of the maggot under such circumstances indicates the septic tank or sewage line is not working properly.

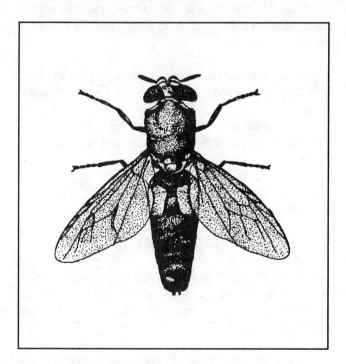

Figure 6. Soldier fly.

VINEGAR (FRUIT) FLY

Fruit flies are nuisance pests and contaminators of food. Fruit flies usually breed in fruit, dirty garbage containers, or slime in drains.

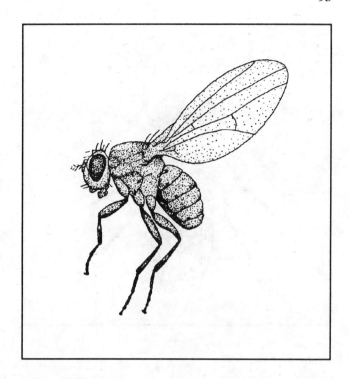

Figure 7. Fruit fly.

Each adult lays about 500 eggs, which hatch and the larvae mature to adults in 9-12 days. These flies are readily attracted to fruit, vegetables, and soda bottles and cans.

EYE GNATS

Eye gnats are small flies, shiny black about 1/6 inch in length. The adults are strongly attracted to moisture around the eyes and noses of people outdoors. In mid-summer large numbers of eye gnats persistently hover around the face causing considerable annoyance. Often eye gnats enter homes through open doors or windows. They are strongly attracted to light and observed mainly on windows in large numbers.

Eye gnats breed in soil containing considerable organic matter, decaying vegetation and animal excrement. The life cycle varies from 7-28 days depending on the temperature and moisture.

Control of eye gnats is difficult. It is often hard to destroy sites over the many acres of land where eye gnats breed. For personal protection outdoor repellents may be applied to the skin. Indoors, space sprays kill adults.

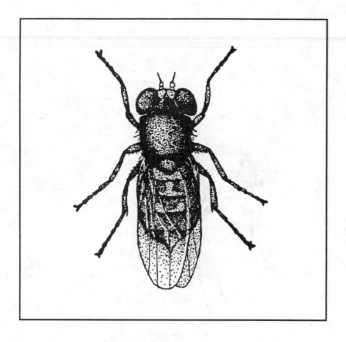

Figure 8. Eye gnat.

Control of Filth Flies

Regardless of advancements in chemical control, sanitation is still the best method of controlling filth flies in and around the home and on the farm. Flies seek breeding places where garbage, animal droppings or vegetation residues accumulate. Locate and thoroughly clean such places. Dry, spread or somehow dispose of dog, cat, or other animal excrement. Do not let garbage accumulate in the open and make sure garbage cans have sound bottoms and tight fitting lids.

Good fitting screens on windows and doors are essential in barring flies from homes, dairy barns, milk rooms, and food-processing areas. Try to make all screen doors open outward. In areas with high humidity such as Florida, copper, aluminum or plastic screens last longer. Galvanized screens deteriorate rapidly.

To kill flies inside the home, use a space spray or aerosol. Release the mist from the aerosol for a few seconds around the room and keep the room closed for 10 to 15 minutes. Outside the house apply a residual or surface spray. Follow dosage and application directions on the container label.

To kill flies in and around farm buildings apply a residual spray, an insecticide bait or a larvicide to the breeding areas.

MOTH FLIES

Moth flies are 1/16 to 1/18 in in length and light gray to tan in color. Their life cycle is 7 to 20 days.

Moth flies breed in decomposing organic material, such as moist plant litter, garbage, sewage, around kitchen or bathroom sinks and water traps in plumbing fixtures.

Non-chemical control is obtained by sanitation and destruction of breeding sites, tight fitting garbage containers and moisture control. For a chemical control you may use larvicides and residual and space sprays.

HUMP-BACKED FLIES

Hump-backed flies are about 1/8 inch long with a small head and large thorax, giving the fly a hump-backed appearance.

Hump-backed flies breed in decaying vegetation, animal debris, garbage, and in ant and termite nests.

Non-chemical control can be obtained by sanitation, destruction of breeding sites and moisture control.

Hump-backed flies may be chemically controlled by residual and space sprays.

DUMP FLIES

Dump flies are about 1/4 inch long, black in color with a shiny thorax and abdomen. They breed in garbage and fowl excrement. The larvae eat the larvae of other flies.

Non-chemical control may be obtained by sanitation and destruction of breeding sites. Larvicides and residual and space sprays may be used for chemical control.

SECONDARY SCREWWORM FLIES

Secondary screwworm flies are about 1/4 inch long, green in color with shiny abdomen and thorax. The latter is marked lengthwise with three dark stripes. Their life cycle is 9 to 39 days. These flies breed in dead tissue surrounding wounds in animals, fresh cut meats and garbage.

Secondary screwworm flies can be non-chemically controlled by sanitation and destruction of breeding sites. Chemical control may be obtained by residual and space sprays.

BLIND MOSQUITOES

Blind mosquitoes are mosquito-like insects in the family Chironomidae. They are often referred to as aquatic midges since their immatures (larvae and pupae) live in water. Blind mosquito is a layman's term which may refer to several species of these midges. Blind mosquitoes do not bite, suck blood, or carry disease. Their occurrence and survival in certain polluted waters often indicates pollution of aquatic habitats. They are important to man only when they emerge in such large numbers that they are a nuisance.

Life Cycle

Figure 9 depicts the life cycle of the blind mosquito. There are 4 stages in the life cycle - egg, larva, pupa and adult. The eggs are laid in a mass on the surface of the water containing 10 to 3,000 eggs depending on the species. Each mass of eggs is enclosed in a gelatinous substance which is usually attached to the edge of the lake, stream or river, and twigs in contact with the water. Egg masses not attached to objects will sink to the bottom where the eggs hatch. Eggs of aquatic midges usually hatch in 2 to 7 days. The newly hatched larvae feed on the gelatinous material for about 2 days.

On the second or third day after hatching, the larvae leave the mass, burrow into the mud or available organic matter or bind with their salivary secretions small inorganic or organic substrate particles to build small tubes and tunnels in which they live. The tubes may also be composed of silk-like threads. Most larval tubes have an opening at each end to allow the larva to feed from either end. Larvae of the burrowing type may live in tubes or tunnels having only one open end. The larva spends most of its time undulating rapidly within the tube to circulate water. From the water the larvae extract oxygen and food. The larvae feed on suspended matter in the water and organic matter in the mud.

After the first molt the larvae of most aquatic midges take on a pink color which gradually darkens into a deep red (some are consequently called blood-worms). As the larvae grow, they enlarge the tube periodically to accommodate their increasing size. The larval stage can last from less than 2 to 7 weeks depending on the water temperature.

The larvae transform into pupae while still in the tubes. The pupal stage normally lasts 3 days. The pupae leave the tube and actively swim to the surface a few hours before the adult emerges.

The adults which emerge mate during swarming at night. The adults do not feed during their adult existence and consequently only live for 3 to 5 days. The entire live cycle can be completed in 2 weeks, although it is common for the life cycle to take longer to complete.

Breeding Sites

Blind mosquitoes are one of the most common and abundant organisms in natural and man-made water systems. In Florida the larvae are abundant in small and large natural lakes, waste water channels, sewage oxidation and settling ponds, and residential-recreational lakes.

Surveys of larval infestations in central Florida has revealed larval populations of $4500/ft^2$ on the bottom of certain lakes. It is the midges emerging from these breeding areas that cause a variety of nuisance and economic problems.

The problem blind mosquitoes in Florida are *Glyptotendipes paripes*, *Chironomus crassicaudatus*, and *Chironomus decorus*, *Goeldichironomus holoprasinus* as well as certain species of *Tanytarsus*. These species usually breed in polluted water 3-12 feet deep.

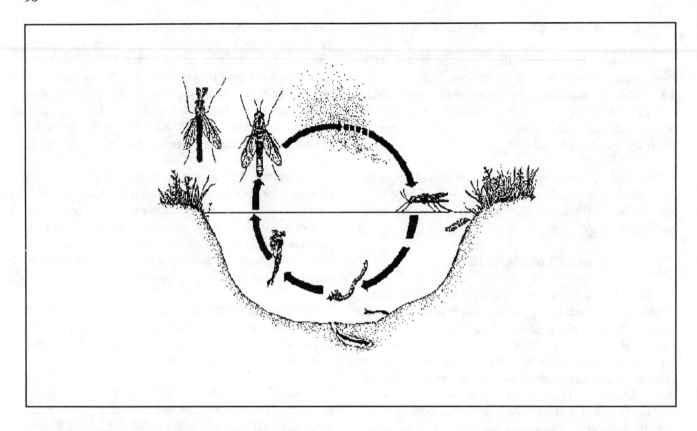

Figure 9. Blind Mosquito Life Cycle (Clockwise from mid-right: egg mass, larva, pupa, adults - male, left; female, right).

Importance

The importance of blind mosquitoes as pests has increased during the past 20-30 years due to:
1. Creation of new midge producing habitats close to residences.
2. Deteriorating water quality which is more suitable for breeding midges.
3. Increasing desire of humans to live close to lakes and rivers.

Residents close to blind mosquito breeding areas experience severe nuisance and economic problems. Blind mosquitoes can emerge in phenomenal numbers between April and November. Often humans have to cease outdoor activity since the adult midges can be inhaled or fly into the mouth, eyes, or ears.

During hot, summer days, midges fly to cool shady places. At night they are attracted to lights around houses and businesses. When large numbers are present, they stain paint, stucco and other wall finishes. Automobiles become soiled, and when headlights and windshields get covered with dead

midges, they can cause traffic accidents. Thousands of dollars are spent on cleaning up the dead bodies of midges from sidewalks and porches the morning after mating flights. Piles of dead midges up to 3 feet deep have been measured in front of stores. The bodies which are mashed to painted surfaces cause permanent staining. Also, blind mosquitoes will fly indoors as doors are opened and closed. Problems indoors such as ruining laundry and staining indoor walls, ceilings, draperies and other furnishings cause severe annoyance for residents.

Where midges are prevalent, spider webs and spiders abound. Accumulations of dead midges and webs require residents to frequently wash and maintain homes and businesses. The dead midges have a smell similar to rotting fish as they decay. The smell persists in damp weather, even after the insects have been removed.

A recent economic impact study undertaken by the Greater Sanford Chamber of Commerce, Seminole County, revealed that blind mosquitoes emerging from Lake Monroe and other nearby bodies of water cause of 3-4 million dollars business loss annually. One lakefront establishment, the Holiday Inn, spends $50,000 each year on property maintenance and blind mosquito control. The same study indicated that at least 10 counties in Florida are affected by similar problems.

Blind mosquitoes are an important component of the food chain in a lake. Fish utilize the larvae as food. Lakes where aquatic midges breed are often our best fishing lakes.

Control

Extensive research has been carried out on the use of insecticides against the larvae and adults of blind mosquitoes. Since the larvae live on the lake or river bottom, they are more difficult to kill than the biting mosquito larvae. The entire water volume must be treated with insecticide to provide effective control. In the past; this total treatment in many instances has been done in small lakes; however, today with emphasis on environmental quality and the development of resistance in midges to pesticides, larval control is not feasible.

Control measures against adult blind mosquitoes are effective for short periods of time. Mists or fogs from boat-mounted or truck-mounted sprayers traveling close to the shoreline kill midges resting in grass or other vegetation near the water's edge before they fly to the buildings. Area control of adult blind mosquitoes should be carried out by organized mosquito control districts. Individuals can kill blind mosquito adults by using fogging units (several attach to lawn mowers or tractors). Follow directions on the label and fogging attachment for application and formulation instructions.

All of these control methods are strictly temporary and do not get to the root of the problem. Blind mosquitoes breed in lakes and rivers in large numbers mainly due to the pollution of the water. Indications are that effluents from food-processing plants, septic tanks, sewage treatment plants, and leaching of fertilizers from lawns and agriculture around lakes, apply nutrients which contribute to the production of food for blind mosquitoes. As pollution increases, the available food increases and blind mosquito populations explode. Blind mosquitoes have been known for years to be indicators of pollution in waterways. But certain lakes in Florida have become so severely polluted that even blind mosquitoes cannot survive in them.

Blind mosquitoes have predators, diseases and parasites which are being investigated as biological control agents. It is hoped that the propagation and establishment of these disease causing organisms parasites, and predators will be a future solution to the blind mosquito problem.

Consequently three long term solutions to the control of blind mosquitoes may be possible: (1) reduce effluents which provide food for the pest or (2) increase the effluents until the pest cannot survive or (3) biological control.

PESTS	DESCRIPTION	BREEDING HABITS	LIFE CYCLE	CONTROL	
				NON-CHEMICAL	CHEMICAL
House Fly	About 1/4" in length; dull gray in color; thorax marked longitudinally with 4 dark stripes; abdomen pale and fourth wing vein is angled.	Warm organic material such as animal and poultry manure, garbage, decaying vegetables and fruits and in piles of moist leaves and lawn clippings.	6 to 10 days	Sanitation and destruction of breeding sites; tight fitting garbage containers and screens on windows and doors.	Larvicides; residual and space sprays; baits.
Green Bottle Fly	About 1/2" in length with shiny metallic blue-green abdomen and thorax.	Dead animal carcasses; decomposing meat and fish; garbage and sometimes animal excrement.	9 to 21 days	Sanitation and destruction of breeding sites; tight fitting garbage containers and screens on windows and doors.	Larvicides; residual and space sprays.
Bronze Bottle Fly	About 1/2" in length with a shiny bronze abdomen and thorax.	Dead animal carcasses; decomposing meat and fish, garbage and sometimes animal excrement.	9 to 10 days	Sanitation and destruction of breeding sites; tight fitting garbage containers.	Larvicides; residual and space sprays.
Blue Bottle Fly	About 1/2" in length with a dull thorax and shiny blue abdomen.	Dead animal carcasses; decomposing meat and fish, garbage, over-ripe fruit, decaying vegetable matter and sores on living humans.	15 to 20 days	Sanitation and destruction of breeding sites; tight fitting garbage containers and screens on windows and doors.	Larvicides; residual and space sprays.
Moth Flies	1/16" to 1/8" in length; light gray to tan in color.	Decomposing organic material such as moist plant litter, garbage, sewage, around kitchen or bathroom sinks and water traps in plumbing fixtures.	7 to 20 days	Sanitation and destruction of breeding sites, tight fitting garbage containers and moisture control.	Larvicides; residual and space sprays.
Eye Gnats	About 1/16" in length, shiny black to dull gray in color with yellow or orange markings on the legs and have small mouthparts and bare aristae on antennae.	Loose soil containing considerable organic material, decaying vegetation and animal excrement.	7 to 28 days	Sanitation and destruction of breeding sites.	Repellents; residual and space sprays.
Hump-backed Flies	About 1/8" in length with a small head and large thorax causing the fly to have a hump-backed appearance.	Decaying vegetation, animal debris, garbage and in ant and termite nests.		Sanitation, destruction of breeding sites and moisture control.	Residual and space sprays.
Blind Mosquitoes	3/16" to 1/2" in length with a slender body, large thorax, small head, and slender 5 to 14 segmented antennae.	Lakes, ponds, rivers, streams, reservoirs and tanks.		Moisture Control	No Suitable Control.
Black Blow Flies	About 1/3" in length with a shiny metallic dark blue abdomen and thorax.	Dead animal carcasses; garbage and wounds in living animals.	10 to 25 days	Sanitation and destruction of breeding sites; tight fitting garbage containers.	Larvicides; residual and space sprays.

PESTS	DESCRIPTION	BREEDING HABITS	LIFE CYCLE	CONTROL NON-CHEMICAL	CHEMICAL
Flesh Flies	3/8" to 9/16" in length, dull grayish-black in color with three dark stripes on the thorax.	Decayed flesh and spoiling meat, manure, wounds in living animals and sometimes garbage.	8 to 21 days	Sanitation and destruction of breeding sites.	Residual and space sprays.
Vinegar Flies	About 1/8" in length, brownish-black to brownish-yellow in color and have a feathery bristle on the antennae.	Fermenting or rotting fruit and vegetable material and in garbage cans.	8 to 10 days	Sanitation and destruction of breeding sites; tight fitting garbage containers.	Residual and space sprays.
Dump Flies	About 1/4" long, black in color with a shiny thorax and abdomen.	Garbage and fowl excrement. The larvae are predaceous on larvae of other flies.		Sanitation and destruction of breeding sites.	Larvicides; residual and space sprays.
Secondary Screwworm Flies	About 1/4" long, green in color, shiny abdomen and thorax with thorax marked lengthwise with 3 dark stripes.	Dead tissue surrounding wounds in animals, fresh cut meats, and garbage.	9 to 39 days	Sanitation and destruction of breeding sites.	Residual and space sprays.

Stored-Food Pests

INTRODUCTION

There are many species of stored-food pests found at various times in food pantries and cabinets. Most of these pests are introduced into our homes in infested food. Some invade homes through normal methods of entry such as open doors.

There is hardly a food item in the kitchen or pantry that can escape being infested by some pest if it remains unused and exposed in some dark corner or drawer long enough. All items are susceptible, including spices, hot pepper, ice cream cones, and even cereals, grits, and cake mixes. Dried flower arrangements and stuffed furniture and toys, though not foods, often harbor infestations.

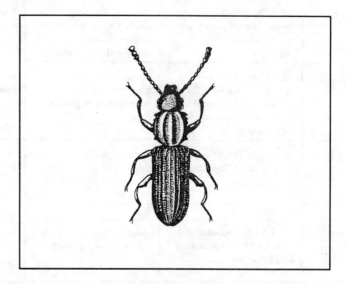

Figure 2. Sawtoothed Grain Beetle.

Some common pantry pests are flour beetles, saw-toothed grain beetles, cigarette beetles, drugstore beetles, larder beetles, granary weevils, rice weevils, spider beetles, grain moths, flour moths, psocids and grain mites. Although there are many different kinds of insects that attack stored food, the damage they produce and the control procedures are similar.

Figure 1. Angoumois Grain Moth.

Stored food pests are economically important and responsible for millions of dollars of loss every year in stored foods and other products. In Florida, most of the stored-food pests can reproduce quickly and have several to many generations in a year. Many species are active the year around under suitable conditions.

Figure 3. Mediterranean Flour Moth.

Large populations of these food pests may develop in unused or undisturbed foods which were infested when purchased. The unwary person leaving a food package open after use can also lead to infestation. From the infested food packages the pests may spread to other exposed food. Often the only way a person knows of their presence is discovery of an infested and infrequently used food item or the encounter of flying or crawling adults coming from storage shelves. Moths and beetles are often attracted to lights or windows and may indicate an infestation. The presence of stored-food pests is not an indication of uncleanliness because infestation may be brought home in purchased food.

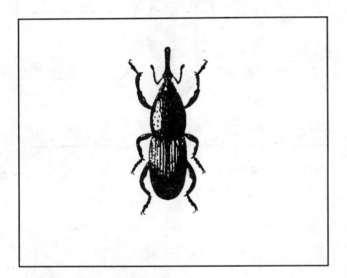

Figure 4. Granary weevil.

CONTROL

The basic fundamentals of controlling stored-food pests are exclusion, chemical treatment, regular inspections, and shelf cleaning. Sometimes an infestation can develop on bread or cracker fragments or in some undisturbed, loose flour or meal on one corner of the shelf. Keep all containers tightly closed. Put susceptible items in tight containers, screw-top jars or other sealable containers. Do not overstock shelves with products that will not be used frequently or soon.

The first step in controlling pantry pests is to locate the source of infestation. All susceptible foods should be thoroughly inspected and badly infested items discarded. Insects in infested foods may be killed by heat or cold. Infested foods may be placed in an oven at 130°F for 30 minutes or a freezer at

0°F. for 4 days to kill insects. If placed in a freezer, the commodity should be used as soon as possible since defrosting usually causes excess moisture in the item. The excess moisture could cause mold later.

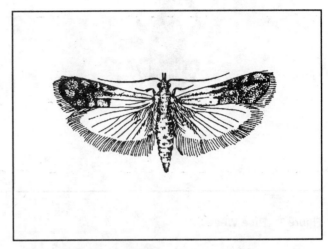

Figure 5. Indian Meal Moth.

To kill insects in the storage area, cabinets and shelves should be thoroughly vacuumed to pick up loose infested material and washed with warm, soapy water. After the storage area is dry, apply a residual spray to the cracks, corners and surfaces of the shelves. Use caution by covering all food items that may be contaminated with spray. Allow the spray application to dry before replacing the food items. Aerosol sprays are helpful in controlling infestations but do not offer effective residual control. Be sure to remove and spray all drawers as well as the inside of the cabinets.

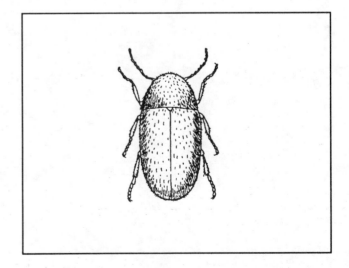

Figure 6. Cigarette Beetle.

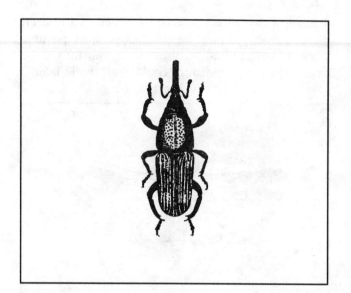

Figure 7. Rice Weevil.

Storage of grains and grain products has resulted in considerable problems with stored-grain insects in homes and institutions. Whole grain can be fumigated in storage containers with carbon dioxide. Dry ice (frozen carbon dioxide) can be placed in storage containers directly on top of grain at the rate of 1/4 lb per 5 gal container. The container lids should not be tightened for 1/2 hour. After 1/2 hour tighten lids and place the commodity in storage. At least 14 days' exposure is necessary to be certain of satisfactory control.

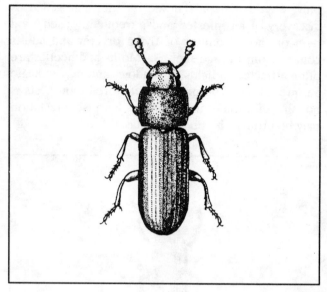

Figure 8. Red Flour Beetle.

Ants

INTRODUCTION

Ants are pests around the home because they feed on and contaminate human foods, infest structures, and build unsightly mounds in lawns. In some cases, ants are able to inflict painful bites or stings. Ants do not attack or eat fabrics, leather or wood in houses; however, some species can establish nests in decaying wood in structures.

Several species of ants are found in or around houses in Florida. In general, the most common ants can be grouped as house-infesting ants, yard-infesting ants, and carpenter ants. The most commonly encountered pest ants are pharaoh, ghost, carpenter, native fire, imported fire, crazy, thief, bigheaded, and acrobat ants.

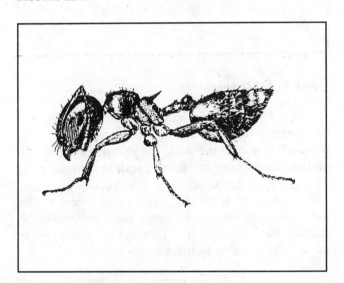

Figure 1. Acrobat ant.

IDENTIFICATION

Ants can be recognized from other insects because they have a narrow waist (pedicel) with one or two joints (nodes) between the thorax and abdomen. Also, ants have elbowed antennae. Winged reproductives have four wings with the first pair being much larger in size than the hind pair.

Ants are frequently confused with termites. However, termites have a broad waist between the thorax and the abdomen. Termite reproductives have four wings of equal size.

Figure 2. Bigheaded ant.

BIOLOGY

Ants are social insects. Three castes (workers, queens, and males) can be found in most colonies. Worker ants, which are sterile females, are seldom winged. They often are extremely variable in size and appearance within a given species (monomorphic - one form; dimorphic - two forms; polymorphic - many forms). The function of the worker is to construct, repair, and defend the nest; and feed the immature and adult ants of the colony, including the queen, and are for the brood.

Queens normally have wings but lose them after mating. The primary function of the queen is reproduction; however, in some of the more highly specialized ants the queen cares for and feeds the first brood of workers on her salivary secretions. The queen may live for many years and is usually replaced by a daughter queen. Depending on the species, ants can have one or more queens.

The male is usually winged and retains its wings until death. The sole function of the male is to mate with an unfertilized female reproductive. After mating occurs, the male dies. Males are produced in old or very large colonies where there is an abundance of food. After reaching maturity, the male usually doesn't remain in the colony very long.

Ants have an egg, larva, pupa, and adult stage. Eggs are almost microscopic in size and hatch into soft legless larvae. Larvae are fed by workers on predigested, regurgitated food. Most larvae are fed liquids, although some older larvae are able to chew and swallow solids. The pupa resembles the adult except that it is soft, uncolored and immobile. In many ant species the pupa is in a cocoon spun by the larva. Six weeks to 2 months are required for development from egg to adult in some species.

Ants establish new colonies by two main methods: flights of winged reproductives and budding. The most common method is for male and female reproductives to leave the nest on mating flights. The mated queen constructs a cavity or cell and rears a brood unaided by workers. The small first brood workers then forage for food. The colony grows in size and numbers as more young are produced.

Budding occurs when one or more queens leave the nest accompanied by workers who aid in establishing and caring for the new colony. Some of the most difficult ant species to control spread colonies by budding. Pharaoh ants, some kinds of fire ants, ghost ants, and Argentine ants spread colonies by budding.

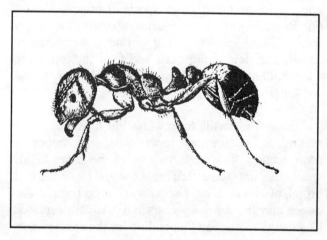

Figure 3. Imported fire ant.

FOOD PREFERENCES

Most ants eat a wide variety of foods, although some have specialized tastes. Fire ants feed on honeydew, sugars, proteins, oils, seeds, plants and insects. Pharaoh ants feed on sugars, proteins, oils and insects. Crazy ants like sugars, protein, and insects; carpenter ants prefer sugars and insects.

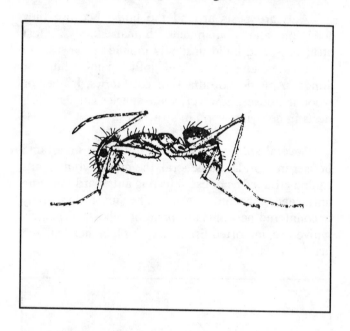

Figure 4. Crazy ant.

Ants locate food by random searching. When a scouting ant finds promising food, she carries it or a piece of it back to the nest. Any workers she meets on the way become excited and rush toward the nest. The means of communication is unknown, but some ants leave scent trails that others can follow to the food source. Ants require water and will travel some distance for it if necessary. Workers are able to bring water to the colony in their stomachs.

CONTROL

Prevention

The best approach to ant control in the home is cleanliness. Any type of food or food particles can attract and provide food for ants. Store food in tight containers. Remove plants that can attract ants or control aphids, whiteflies and other insects that produce honeydew. Reduce moisture sources, including condensation and leaks.

Inspection

Location of the nest is the key to control because ants are social insects. Large numbers of individual ants can be killed without ever solving the problem. Determine the kind of ant species. Most species of ants never enter buildings; others build their nests near buildings and forage indoors. Others usually nest indoors.

Keep a record of where ants have been seen. Some ants follow definite trails. If possible, follow these trails to the nest. Placement of attractive materials, such as jelly, oils, protein and other materials can attract large numbers of ants so they can be followed to their nest.

Often children like to watch ants and can be very useful in tracing their trails. Outdoors, ant nests can often be located by seeing ant hills on the ground. Some ants deposit earth on the soil surface when they construct the nest. Fire ants and certain other ants build conspicuous mounds. Nests may also be constructed next to or under the house foundation, under sidewalks, driveways and patios, or in decaying logs or tree trunks.

Indoors, ants may nest in walls, behind a baseboard or under the house. Often ant trails enter through a crack but the nest may be some distance away. Some ants may also nest in decayed or rotting wood in the house.

Chemical Control

Chemical control of ants can be applied as barrier, drench, nest and bait treatments. Insecticides, sprays, dusts, granules and baits are useful in ant control.

Barrier treatment prevents outdoor nesting species from foraging indoors or cuts off an indoor nest from food and water sources. Barrier treatments should be applied to baseboards, door and window frames, around chimneys, cracks and crevices, around sinks and toilets, and between walls and flooring. Treat areas where ants have trails and move into or through the house. Sprays can be applied as barrier treatments. Microencapsulated (for commercial pest control operators) or wettable powder formulations usually are the best formulations to use because they

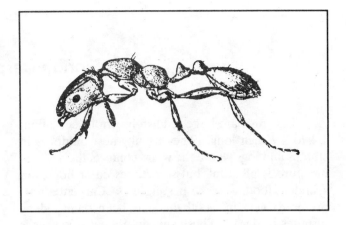

Figure 5. Pharaoh ant.

can be tracked back to the nest by returning workers. Barrier treatments can cause colony splitting of species that reproduce by budding (for instance, Pharaoh, Argentine, and ghost ants).

Drench treatments are useful for the control of mound-dwelling species outdoors. For instance, fire ant colonies near homes can be destroyed by pouring drenches on the mounds.

Nest treatments are the most effective ant control. Apply an insecticide to the nest. If the nest cannot be located, pay particular attention to places where ants enter the house or room. Dusts are best for this type of control because they can be puffed into wall voids that contain ant nests. Dusts provide the longest residual control in dry areas and can be tracked into the nest by foraging ants.

Bait treatments are effective for control of many ant species if they eat the bait. Ant baits are labeled for residential and commercial use. They are enclosed in a childproof plastic tray and are broadly labeled for many ant species.

Boric acid is a good ingredient for ant baits and used in several commercial products. A liquid bait can be made with the following formula: Mix 1 level teaspoon of boric acid in 2 1/2 fl. oz. of corn syrup or honey. Heat until boric acid completely dissolves. Cool. Dilute bait with an equal volume of water and with an eye-dropper, place in areas where ants feed, or in small lids where ants have access. Be sure to keep baits available to ants for 2 weeks. Keep bait mixture out of reach of children.

Florida Carpenter Ants

These ants are large reddish-brown insects about 1/4 to 1/2 inch long. They usually nest outdoors in stumps and logs where the wood contacts the soil and moisture is plentiful, but sometimes enter homes in search of food, water or nesting sites. Carpenter ants prefer to nest in wood that has been damaged by termites or decay. These ants do not eat wood (as is the case with termites) but excavate galleries in it to rear their young. They feed on honeydew from suckling insects and household food scraps and do not damage sound wood to any extent. They eject the wood in the form of a coarse sawdust. Carpenter ant galleries are kept smooth and clean and have a sandpapered appearance. Other wood infesting insects do not keep their galleries clean.

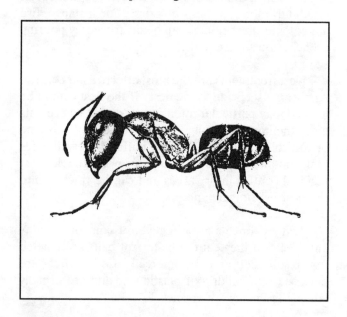

Figure 1. Carpenter ant.

CONTROL

Prevention

The best approach to ant control in the home is cleanliness. Any type of food or food particles can attract and provide food for ants. Store food in tight containers. Remove plants that can attract ants or control aphids, whiteflies, and other insects that produce honeydew. Reduce moisture sources, including condensation and leaks.

Inspection

Location of the nest is the key to control because ants are social insects. Large numbers of individual ants can be killed without ever solving the problem. Determine the kind of ant species. Most species of ants never enter buildings; others build their nest near buildings and forage indoors. Others usually nest indoors.

Keep a record of where ants have been seen. Some ants follow definite trails. If possible, follow these trails to the nest. Placement of attractive materials, such as jelly, oils, protein, and other materials can attract large numbers of ants so they can be followed to their nest.

Find the ant nest by watching the movements of ants. Often children like to watch ants and can be very useful in tracing their trails. Outdoors, ant nests can often be located by seeing ant hills on the ground. Some ants deposit earth on the soil surface when they construct the nest. Fire ants and certain other ants build conspicuous mounds. Nests may also be constructed next to or under the house foundation, under sidewalks, driveways and patios, or in decaying logs or tree trunks.

Indoors, ants may nest in walls, behind a baseboard, or under the house. Often ant trails enter or walk through a crack but the nest may be some distance away. Some ants may also nest in decayed or rotting wood in the house.

Chemical Control

Chemical control of ants can be applied as barrier treatments, drench treatments, nest treatments, and bait treatments. Sprays, dusts, granules, and baits are formulations of insecticides useful in ant control.

The purpose of a barrier treatment is to prevent outdoor nesting species from foraging indoors or cut off an indoor nest from food and water sources. Barrier treatments should be applied to baseboards, door and window frames, around chimneys, cracks and crevices, around sinks and toilets, and between walls and flooring. Treat areas where ants have trails and move into or through the house. Sprays can be applied as barrier treatments. Microencapsulated (for commercial pest control operators) or wettable powder formulations usually are the best formulations to use because they can be tracked back to the nest by returning workers. Barrier treatments can cause colony splitting of species that reproduce by budding (for instance, Pharaoh ant, Argentine ant, and ghost ant).

Drench treatments are useful for the control of mound dwelling species outdoors. For instance, fire ant colonies near homes can be destroyed by pouring drenches on the mounds.

Nest treatments are the most effective way of controlling ants. If the nest is located, apply an insecticide to the nest. If the nest cannot be located, pay particular attention to places where ants are gaining access to the house or room. Dusts are thebest formulation for this type of control because they can be puffed into wall voids that contain ant nests. They will provide the longest residual control in dry areas and can be tracked into the nest by foraging ants.

Bait treatments are effective for control of many ant species if the baits are consumed by the ants. Ant baits are labeled for residential and commercial use. They are enclosed in a childproof plastic tray and are broadly labeled for many ant species.

Boric acid is a good ingredient for ant baits and is used in several commercial products. A liquid bait can be made with the following formula: Mix 1 level teaspoon of boric acid in 2 1/2 fl. oz. of corn syrup or honey. Heat until boric acid completely dissolves. Cool. Dilute bait with an equal volume of water and with an eye-dropper place in areas where ants feed, or place in small lids where ants have access. Be sure to keep baits available to ants for 2 weeks. Keep bait mixture out of reach of children.

Imported Fire Ants

The imported fire ant is a small, aggressive ant that builds a rounded, somewhat conical nest or mound, often 2 or 3 feet across. However, in sandy soil the mound does not maintain its shape. In Florida, the imported fire ant infests all counties. The total area infested in Florida is estimated at over 30 million acres.

HISTORY AND REGULATORY PROGRAMS

There are two species of imported fire ants. The black imported fire ant, *Solenopsis richteri*, was imported to the United States in 1918 or earlier. This ant now occupies only a small area in Alabama and northern Mississippi. The red imported fire ant, *Solenopsis invicta*, was not Imported until about 1940, and has achieved its remarkable spread in only 40 years. This ant presently infests more than 230 million acres in Alabama, Arkansas, Florida, Georgia, Louisiana, Mississippi, North Carolina, South Carolina, and Texas.

Since 1957, cooperative state and federal programs have been conducted to retard the spread of fire ants, to survey the degree of ant infestations, and to provide temporary relief to farmers and residents of heavily infested areas.

Quarantine programs, aimed at preventing the spread of fire ants from Florida to other states, restrict the movement of infested articles. Soil, potted plants, plants with soil attached, grass sod, hay, and used soil moving equipment cannot be moved from Florida to uninfested parts of the country without undergoing inspection. Once inspectors of the Florida Division of Plant Industry certify these items as free from infestation, the articles may be moved to other locations.

In the past, regulatory agencies have been involved with efforts to control or eradicate the red imported fire ant in various areas of the southeastern United States. The pesticides heptachlor and dieldrin were used to control fire ants from 1957 to 1962. Mirex was the pesticide heavily relied upon for fire ant control from 1962 to 1977. However, in 1978, mirex was banned by the Environmental Protection Agency because of mirex residues found in non-target organisms and humans, and the suspicion that it might be carcinogenic.

DESCRIPTION AND BIOLOGY

Imported fire ants are 1/8 to 1/4 inch in length and are reddish brown to black (Figure 1). They are social insects and live in colonies which may have up to 200,000 individuals. Fire ant colonies are made up of a queen ant, winged males and females (virgin queens), workers, and brood (eggs, larvae, and pupae).

Since it is not necessary nor desirable to treat for native fire ants, it is important to know the difference between the native fire ant, *Solenopsis geminata*, and the red imported fire ant. The head of the imported fire ant large worker is not wider than the abdomen, whereas the head of the native fire ant large worker is wider than the abdomen. Imported fire ant nests are rounded and conical; nests of native fire ants are irregular in shape.

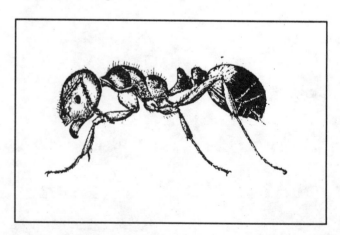

Figure 1. Imported fire ant.

The winged reproductives mainly leave the fire ant colony on mating flights in late spring and summer, although swarming may occur at any time of the year. The ants mate during flight, and the females land to begin a new colony. Most females fly or are blown less than 1 mile from the nest, but some may travel 12 miles or more from the nest. Fire ant nests are normally prevalent in open, sunny areas. Pastures and other farm lands, roadsides, and home yards are areas often infested.

In heavy soils, each mature colony of imported fire ants can construct a mound that is sometimes as much as 2 feet high and 3 feet in diameter. In many heavily infested areas there are as many as 50 mature colonies per acre. Infestations of 20 to 30 colonies per acre are common. In freshly invaded areas, there may be several hundred small, new pests per acre. Imported fire ants achieve their greatest density in high maintenance areas such as lawns, parks, roadsides, pastures, and areas disturbed by flooding, draining, or plowing.

DAMAGE

The imported fire ant causes damage that is difficult to measure in dollars. Its painful, burning sting results in pustules that may take up to ten days to heal. If broken, the pustule may become infected. Some people exhibit an allergic reaction to fire ant stings. Such a reaction may cause paralysis or heart attack. A few individuals have died as a result of allergic responses to fire ant stings, but this is very rare. More people die from bee stings than fire ant stings.

Fire ant mounds in yards, playgrounds, and recreational areas are a hazard to children and pets. On farms, mounds may cause damage to machinery used to harvest crops. In addition, farm laborers may refuse to work on land where ants are prevalent because of the numerous stings that they could receive. Fire ants sometimes reduce the yield of farm crops by feeding on germinating seeds and seedlings. A significant reduction in yield of soybeans has been noted in fields infested with 40 to 50 colonies per acre.

CONTROL

Fire ant control can be achieved by individual mound treatment or by area treatment. Individual mound treatment can be accomplished by several methods. Applications for non-agricultural land are:

* Insecticide mound drenches. Dilute insecticide concentrates in water according to label instructions, and apply (with a sprinkling can) one gallon of solution per mound. Thoroughly wet the mound and surrounding area. Best results are achieved when mound drenches are applied in cool weather (when soil surface temperatures are 60° to 80° F) or after a rainfall. Check treated mounds seven days after application and retreat as necessary.

* Boiling water mound drenches. Heat 1/2 gallon of water to the boiling point. Apply hot water directly to the mound and surface surrounding the fire ant mound when the brood is near the surface of the soil (on cool days after 10:00 a.m. when sun is shining). CAUTION: Boiling water is quite dangerous since it is easily spilled onto the skin. Extreme caution must be taken to prevent burns during application.

* Baits. Apply 5 tablespoons of bait per fire ant mound. Distribute the bait uniformly 3 to 4 feet around the base of the mound. Do not contaminate kitchen utensils during application or storage.

Area treatment may be accomplished by broadcasting bait at 1.0 to 1.5 pounds of formulated bait per acre. Area treatments may be required at least annually because fire ants reinvade treated areas.

Baits are registered for fire ant control in pasture, range grass, lawns, turf, and non-agricultural lands. The formulations are composed of a corn carrier and soybean oil. The soybean oil may become rancid after the bag is opened. Consequently, it is recommended that the bait be used within 3 days of the package being opened. Fire ants do not feed on rancid bait.

An effective fire ant bait must be slow acting, so that it can be passed by the workers throughout the ant colony and eventually to the queen. The baits are slow acting and require 2 to 4 weeks to kill or sterilize the queen. Within 4 to 8 weeks a significant number of ants should die, and activity in the mound should be reduced. Live worker ants may be found in mounds for 4 to 6 months, even though the queen is dead. Retreat these mounds in 4 months if they are a nuisance.

To insure ant activity, all bait treatments for fire ants should be made when soil surface temperatures are above 60°F or after a rainfall. Applications during the hot times of day, when soil surface temperatures are above 80°F, are also undesirable.

COMMENTS ON CONTROL MEASURES

Individual mound treatments should be used when only a few mounds need to be treated, such as in yards and lawns. When portions of a large tract of land have a particularly heavy population, those infested areas should be treated either by air or by ground application. The purpose of the control program is to maintain the fire ant population below a damaging level. Repeat applications may be necessary to keep the fire ant population suppressed.

Proper application of control measures for fire ants will result in suppressing the fire ant populations as well as safeguarding the environment. Remember to read and follow label directions on all pesticides before application. Store them safely out of reach of children.

Cockroaches[1]

P. G. Koehler and J. L. Castner[2]

German cockroach, *Blattella germanica.* This is the most important species of cockroach in the United States. It is about 1/2" to 5/8" long as an adult. Nymphs and adults of both sexes have two dark stripes behind the head. It prefers to live in kitchens and bathrooms of homes and apartments, restaurants, supermarkets, and hospitals. The Asian cockroach, *Blattella asahinai,* is identical to the German cockroach in appearance and lives outdoors in lawns and leaf litter.

Brown-banded cockroach, *Supella longipalpa.* It is about 5/8" long as an adult. This cockroach is dark brown, and the wings range from reddish brown to brown. There are two pale brown bands on the wings, and the edge of the pronotum is clear. It prefers to live in bedrooms, furniture, and closets, particularly high on shelves.

American cockroach, *Periplaneta americana.* This cockroach is about 1 1/2" long as an adult. It is reddish-brown with light markings behind the head. The cerci at the tip of the abdomen are long and thin. It is commonly found in sewers and basements.

Australian cockroach, *Periplaneta australasiae.* It is about 1 1/4" to 1 1/2" long as an adult. It is reddish-brown to dark brown with a characteristic marking behind the head. On the front edge of the base of the forewing, there is a light yellow band. Nymphs have light yellow spots on the top of the abdomen. This cockroach is abundant outdoors and in greenhouses where it can damage plants. It enters homes and is called a palmetto bug.

Smoky-brown cockroach, *Periplaneta fuliginosa.* It is about 1 1/4" long as an adult. It is mahogany brown to black with no patterns behind the head. This cockroach is abundant outdoors and is found in tree holes, wood piles, and attics of houses in Florida. It readily enters homes and is called a palmetto bug.

Brown cockroach, *Periplaneta brunnea.* This cockroach is almost identical to the American cockroach in appearance and is about 1 1/4" long as an adult. It is reddish-brown. The cerci at the tip of the abdomen are stubby; whereas the American cockroach has long, thin cerci. The brown cockroach is found outdoors. It readily enters houses and is often called a palmetto bug.

Florida woods cockroach, *Eurycotis floridana.* It is 1 1/2" to 1 3/4" long as an adult and is often called the stinking cockroach; it produces a foul-smelling fluid to protect it from predation. It is dark reddish-brown to black. The nymphs have broad yellow bands on the top of the thorax. This cockroach is commonly found in leaf mulch, wood piles, and under rotting logs. It is often called a palmetto bug.

Surinam cockroach, *Pycnoscelus surinamensis.* This cockroach is about 3/4" to 1" long. It is shiny brown to black with golden markings on the abdomen. The pronotum behind the head has a yellow margin along the front edge. It is a burrowing species that lives outdoors and often infests potted plants. When plants are brought inside, the cockroach then infests the premises.

Oriental cockroach, *Blatta orientalis.* It is about 1" long as an adult. It is shiny black and has no distinctive markings. The male has wings that cover only about 3/4 of the abdomen; the female has only wing pads or lobes. This cockroach is not commonly found in Florida. It is usually found in damp basements, sewers, and crawl spaces beneath houses.

Cuban cockroach, *Panchlora nivea.* The Cuban cockroach is about 3/4" long as an adult. Males and females are pale green, whereas the nymphs are dark brown. It is an outdoors, tropical species that usually is not found north of Florida. Adults are attracted to light and are adept fliers.

---

1. This document was published as Fact Sheet ENY-501, Florida Cooperative Extension Service, May, 1991. For more information, contact your county Cooperative Extension Service office.

2. Professor and Scientific Photographer, respectively, Department of Entomology and Nematology, Institute of Food and Agricultural Sciences, University of Florida, Gainesville.

German cockroach

Brown-banded cockroach

American cockroach

Australian cockroach

Smoky-brown cockroach

Brown cockroach

Florida woods cockroach

Surinam cockroach

Oriental cockroach

Cuban cockroach

Institute of Food and Agricultural Sciences
Department of Entomology and Nematology

SP 121

Filth-breeding Flies

P. G. Koehler and J. L. Castner[1]

Filth-breeding flies are annoying and important vectors of disease. The larvae usually are found in excrement, decaying vegetable matter, rotting meat, or garbage. They are strong fliers and often enter the house in search of food or moisture. They have sponging-lapping mouthparts and ingest only liquids.

House fly, *Musca domestica*. It is 4 to 6mm long, gray to black in color, and the thorax has four black stripes. The fourth wing vein has a sharp upward bend to almost join the third at the tip of the wing. The fly prefers to breed in decaying animal excrement but will also develop in garbage and vegetable waste. High populations are sometimes found close to poultry and livestock facilities.

Green bottle fly, *Phaenicia sericata*. It is 10mm long and metallic green with no stripes on the thorax. The bristles on top of the thorax are large, and the front of the thorax is covered with whitish pollen dust. The anterior spiracle is dark. It prefers to live on dead animals, garbage, and high protein fecal material. It is prevalent in dog stools.

Blue bottle fly, *Calliphora vomitoria*. It is 10mm long, the thorax is dull, and the abdomen is metallic blue-green. There are no stripes on the thorax. The bristles on top of the thorax are small. The anterior spiracle is red. It breeds in dead animals and garbage.

Secondary screwworm fly, *Cochliomyia macellaria*. It is 10mm long with the thorax and abdomen metallic green and shiny. The thorax has three stripes of equal length. It breeds in decaying animal flesh.

Vinegar fly, *Drosophila melanogaster*. It is 3mm long and can enter houses through normal screening. The fly has a tan-colored head and thorax with a black abdomen. Its larvae feed on yeast in the fermenting fluids of decaying fruit and vegetables. It is common wherever these items ferment and rot.

Flesh fly, *Sarcophaga spp.* It is 8mm long, dull grayish in color with three stripes on the thorax. The abdomen has a checkerboard gray pattern. Many species have a red tip on the abdomen and red eyes. It breeds on decaying meat, fish, and garbage. The fly lays live larvae on meats and scraps.

Moth fly, *Psychoda spp.* It is 2 to 3mm long and is tan to grayish. The wings are without cross veins and mottled with black and white. The antennae are 13 segmented and have a puff of hairs on each segment. It breeds in sewage filter plants and is often called a filter fly. It also breeds in moist areas in the house such as clogged overflow pipes of sinks and tubs; therefore it is often called a drain fly.

Eye gnat, *Hippelates pusio*. It is 2 to 3mm long and is shiny black in color. It breeds in decaying vegetable matter outside. The fly frequently feeds around the eyes and causes irritation and annoyance. Injury to the eye can be caused by spines on the mouthparts.

Humpbacked fly (family Phoridae). It is about 3mm long with a dull-colored, black body. The antennae are one segmented. Wings have three posterior veins with no cross veins. It breeds in decaying organic matter of high protein. Residues in trash cans and drains are the most frequent areas of breeding. It is frequently found in hospitals, nursing homes and supermarkets.

Soldier fly, *Hermatia illucens*. The soldier fly is about 15mm long and is wasp-like in appearance. The abdomen has two whitish areas near the base. It breeds in wet animal feces and commonly invades the house through the toilet from infestations in the septic tank or broken sewer lines.

1. Professor and Scientific Photographer, respectively, Department of Entomology and Nematology, Cooperative Extension Service, Institute of Food and Agricultural Sciences (IFAS), University of Florida, Gainesville.

COOPERATIVE EXTENSION SERVICE, UNIVERSITY OF FLORIDA, INSTITUTE OF FOOD AND AGRICULTURAL SCIENCES, John T. Woeste, Director, in cooperation with United States Department of Agriculture, publishes this information to further the purpose of the May 8 and June 30, 1914 Acts of Congress; and is authorized to provide research, educational information and other services only to individuals and institutions that function without regard to race, color, sex, age, handicap or national origin. Information on copies for purchase is available from C. M. Hinton, Publications Distribution Center, IFAS Building 664, University of Florida, Gainesville, Florida 32611. Printed 11/92.

Filth-breeding Flies

House fly

Green bottle fly

Blue bottle fly

Secondary screwworm fly

Vinegar fly

Flesh fly

Moth fly

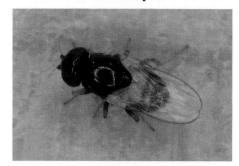

Eye gnat

Humpbacked fly

Soldier fly

Institute of Food and Agricultural Sciences

Department of Entomology and Nematology

SP 128

Stored Product Pests

P. G. Koehler, W. H. Kern and J. L. Castner[1]

Infestations of stored product pests are usually caused by bringing infested food home. Although adults are often found, the larvae cause damage to the commodity. Stored product pests are either beetles or moths and are grouped as pests of whole grain, grain products, beans, and general feeders.

Rice weevil, *Sitophilus oryzae.* It is 2 to 3mm long and reddish brown. The thorax has round pits, and the wing covers have four light spots. It usually attacks whole corn but has been found in macaroni and spaghetti. The life cycle can be completed in 30 days.

Lesser grain borer, *Rhyzopertha dominica.* It is 3mm long and dark brown to black in color. Its head is hidden beneath the thorax. The thorax is very rough, and the overall shape of the beetle is cylindrical. It attacks whole grain. The eggs are laid in clusters on the surface, and the larvae burrow into the kernels. The life cycle takes about 58 days.

Flour beetles, *Tribolium spp.* The two major species of flour beetles are the red and confused flour beetle. They are both reddish brown and 3 to 4mm long. Each antenna of the red flour beetle has a three segmented club; the antenna of the confused flour beetle has a four segmented club. Red flour beetles fly; confused flour beetles do not fly. They infest flour and milled grain. The life cycle takes about 30 days.

Sawtoothed grain beetle, *Oryzaephilus surinamensis.* It is about 2.5mm long and is brown in color. The body is flattened and the thorax has six sawtoothed projections on each side. It is common in cereal products and macaroni. Because of its size, it can enter packaging through tiny cracks and folds. It cannot fly, and its life cycle takes about 30 days.

Cigarette beetle, *Lasioderma serricorne.* It is 3mm long and reddish brown. The head is bent down under the thorax; the antennae do not have a club. The wing cover is smooth and covered with golden hairs. Besides attacking tobacco, it infests spices, seeds, and dog food. The life cycle takes 30 to 50 days.

Drugstore beetle, *Stegobium paniceum.* It is 2.5mm long and brown in color. It has a three-segmented club on the antennae. The wing covers have longitudinal lines. It infests flour, cereal, spices, dog food, and many other products. It has a life cycle of about 60 days.

Weevils (family Bruchidae). The bean weevil is olive brown, mottled with dark brown and gray, and it is 3mm long. It has one large and two small spines on the tip of the hind femur, and the legs and antennae are reddish. The cowpea weevil has a large spot in the middle of the wing cover, with black tips on the wings. These weevils attack stored beans and peas. The life cycle can be completed in 21 to 80 days.

Mealworms, *Tenebrio spp.* Mealworms are the largest insects attacking stored grain and are 13mm long. The yellow and the dark mealworms are the two most common species. They are shiny black with well-developed wings. They fly and are attracted to light. They infest milled grain that has high moisture. The life cycle can be completed in 10 months.

Almond moth, *Cadra cautella.* It has a wing spread of about 15mm. The wings are brownish or blackish gray with the base lighter than the wing tip, which also has a fringe of hairs. The almond moth infests dried fruit and nuts. The life cycle can be completed in 60 days.

Indianmeal moth, *Plodia interpunctella.* It has a wingspread of about 19mm. The wings are tan on the basal one third and coppery colored on the rest. It spins webs on the infested product. The life cycle can be completed in 60 days.

1. Professor, Research Assistant and Scientific Photographer, respectively, Department of Entomology and Nematology, Cooperative Extension Service, Institute of Food and Agricultural Sciences (IFAS), University of Florida, Gainesville.

Stored Product Pests

Rice weevil

Lesser grain borer

Red flour beetle

Sawtoothed grain beetle

Cigarette beetle

Drugstore beetle

Cowpea weevil

Mealworm adult

Almond moth

Indianmeal moth

Institute of Food and Agricultural Sciences
Department of Entomology and Nematology

SP 119

Pest Ants

P. G. Koehler, K. M. Vail and J. L. Castner[1]

Acrobat ant, *Crematogaster spp.* It is usually light brown to shiny black and 3mm long. There are two nodes on the petiole between the thorax and abdomen. The abdomen is heart shaped. The antennae are 12 segmented and have a two-segmented club. There are two spines on the thorax. Workers are monomorphic (one form), and colonies are polygyne (many queens). Acrobat ants hold their abdomens over their heads when disturbed.

Argentine ant, *Linepithema humile.* It is about 2 to 3mm long and dark brown. It has one node on the petiole. The scape (first segment of the antenna) is not longer than its head. There is no circle of hairs at the tip of the abdomen. Workers are monomorphic, and colonies are polygyne.

Bigheaded ant, *Pheidole megacephala.* It is yellowish brown and 2 to 3mm long. There are two nodes on the petiole. Antennae are 12 segmented with a three-segmented club. There are two spines that project from the thorax towards the abdomen. Workers are dimorphic (two sizes—major and minor workers), and colonies are polygyne. Major workers have very large heads.

Crazy ant, *Paratrechina longicornis.* It is black and 3mm long, with long legs. It has one node on the petiole. The antennae are 12 segmented with no club, and the scape is two times the length of the head. There is a circle of hairs at the tip of the abdomen. Workers are monomorphic, and colonies are polygyne. Workers move erratically.

Florida carpenter ant, *Camponotus abdominalis floridanus.* It is a large ant, about 5 to 10mm long, with a red thorax and black abdomen. It has one node on the petiole. The scape is about as long as the head. The tip of the abdomen has a circle of hairs. The thorax is evenly rounded when viewed from the side. Workers

are polymorphic (many sizes), and colonies are monogyne (one queen). Carpenter ants nest in hollow voids and pile wood chips nearby.

Ghost ant, *Tapinoma melanocephalum.* This ant is 1mm long, with a black head and thorax and clear abdomen and legs. The petiole has one node and is hidden by the abdomen. Workers are monomorphic, and colonies are polygyne. This ant has a musty odor when it is squashed.

Imported fire ant, *Solenopsis invicta.* It is reddish brown and 3 to 6mm long. There are two nodes on the petiole. The antennae are 10 segmented with a two-segmented club. There are no spines on the thorax; the mandibles have four teeth. Workers are polymorphic, and colonies are usually monogyne but sometimes polygyne. Fire ants build large mounds. Workers can sting, commonly causing a white pustule to form.

Little fire ant, *Wasmania auropunctata.* It is golden brown and 1 to 2mm long. There are two nodes on the petiole and one pair of spines on the thorax. The antennae are 11 segmented and the last segment is long. The head is covered with grooves. Workers are monomorphic, and colonies are polygyne.

Native fire ant, *Solenopsis geminata.* It is reddish brown to black and 3 to 6mm long. There are two nodes on the petiole. The antennae are 10 segmented with a two-segmented club. There are no spines on the thorax, and the mandibles have no teeth. Workers are polymorphic and the major workers have a large head (wider than the thorax). Colonies are monogyne. Workers can sting, but no white pustule forms.

Pharaoh ant, *Monomorium pharaonis.* It is rust colored and 2mm long. There are two nodes on the petiole. Antennae are 12 segmented with a three-segmented club; no spines on the thorax. Workers are monomorphic, and colonies are polygyne.

1. Professor, Research Assistant and Scientific Photographer, respectively, Department of Entomology and Nematology, Cooperative Extension Service, Institute of Food and Agricultural Sciences (IFAS), University of Florida, Gainesville.

Pest Ants

Acrobat ant

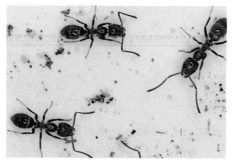

Argentine ant

Bigheaded ant

Crazy ant

Florida carpenter ant

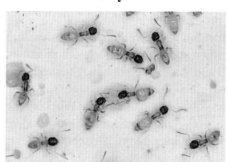

Ghost ant

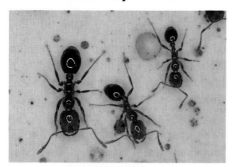

Imported fire ant

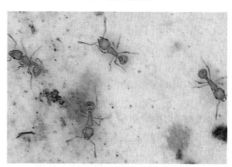

Little fire ant

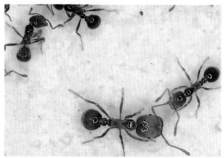

Native fire ant

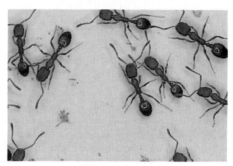

Pharaoh ant

Fabric and
Wood-Destroying
Pests

Carpet Beetles

CARPET AND HIDE BEETLES

Carpet beetles can damage fabrics, furnishings and clothing that contain natural animal fibers such as wool, silk, hair, bristles, fur or feathers. Synthetic items are resistant to attack, but mixtures of synthetic and natural fibers can be damaged. The natural habitats of carpet beetles are nests of birds, rodents, insects and spiders. They then can spread into homes to damage carpets, rugs and clothing. They also may feed on pollen and can be carried into the house on cut flowers.

Several species of carpet beetles are of importance. Hide and larder beetles prefer to attack hides and skins although they also like high protein products such as ham, bacon and cheese. The black carpet, common carpet, furniture carpet and varied carpet beetles usually infest fabrics and carpets.

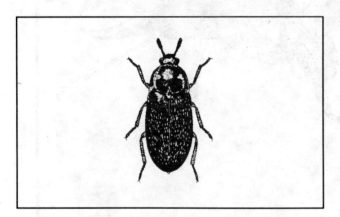

Figure 1. Hide beetle.

Identification

Carpet beetles vary considerably in size and appearance, but their habits, damage and control methods are similar. The adults are hard-shelled oval beetles about 1/8 inch in length. Black carpet beetles are usually black with brown legs. Adults of the common carpet beetle, furniture carpet beetle and varied carpet beetle are usually mottled white, red, yellow and black, and covered with body scales.

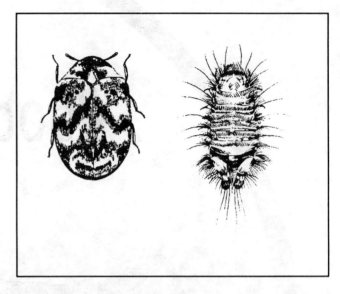

Figure 2. Varied Carpet beetle.

Carpet beetle larvae are tiny when they hatch and have a distinctive carrot shape with tail bristles. Larvae of the black carpet beetle are longer than other species. They are dark brown with a tuft of long brown hair at the end of the body. The larvae of other species are short, stubby and robust, and are covered with brown or black bristles.

Life Cycle

The life cycle of carpet beetles can take from 3 months to 2 years. The adult female beetles lay up to 100 eggs which hatch into larvae in 1 or 2 weeks. The larvae can develop under a wide range of temperature and humidity conditions. They tend to avoid light and actively feed in dark cracks, crevices or folds. Most frequently they are found in closets, drawers or inside upholstered furniture. A larva requires from 9 months to as much as 3 years to complete its growth, depending upon climatic conditions and food. The larvae wander considerably and may be found anywhere in a building. Adults are attracted to light and are often found on windows or around flowers. The adults are good fliers and may spread in materials infested with eggs, larvae or pupae.

Control

Prevention

Good housekeeping prevents infestations. Frequent cleaning of floors to remove dust and lint eliminates much of the available food supply, although dust and lint may collect in inaccessible areas. Clothes should be drycleaned regularly.

Stored materials subject to beetle damage should be thoroughly cleaned before storage. The storage area should be a chest, closet or other container that can be tightly closed and into which moth balls or flakes can be placed at the rate of 1 lb. per 50 cubic feet of space. The initial treatment should kill the carpet beetles; however, this method should not be relied upon to protect stored fabrics over an extended period of several months or more. Stored materials should be periodically sunned and brushed. Cedar chests will not kill carpet beetles.

Inspection

Find sources of carpet beetle infestations. Although carpet beetles damage fabrics in a house, they may also be found in nests built by sparrows, starlings or pigeons on or inside the house. Wasp nests under eaves or in attics are also common sources of carpet beetles because the larvae will feed upon the skins and larvae of wasps. Dead insects in attics or inaccessible areas may also be an important reservoir of infestation. Also, accumulations of debris in ventilation ducts in the house may be a source.

Nonchemical Control

Adult carpet beetles can be captured on sticky fly paper baited with animal products. Cedar products can be used to protect susceptible items. Newly hatched larvae die when exposed to cedar, but older larvae and adults are not affected. The heartwood of red cedar has a vapor that is toxic to larvae, but after cedar is more than 36 months old it is useless for control.

Plastic bags and tight containers can be used to store garments. These containers prevent adult beetles from laying eggs on or near susceptible clothing. But if the clothing is infested, the bags will confine the infestation to just a few items.

Cold storage has been long used to protect articles attacked by carpet beetles. Clothing, coats and sweaters stored at 40-42°F will be protected for long periods of time. Freezing has also been used to kill carpet beetles. Infested materials should be placed in plastic bags and loosely packed in a chest freezer at -20°F for 3 days. Reducing the air in the bag eliminates the formation of ice. Heat has also been used to kill or repel carpet beetle larvae. Exposure of infested items to 105°F for 4 hours is sufficient; placing infested items in the hottest rays of the sun causes larvae to abandon the fabric.

Chemical Control

Several kinds of sprays are registered for control of carpet beetles. Spot treatment with sprays can provide effective control of carpet beetles, although more than one application may be necessary. Dusts may also be effective. When spraying and dusting rugs, closet areas, and storage areas, be sure to apply them to all known or suspected feeding and hiding places.

Clothes Moths And Plaster Bagworms

CLOTHES MOTHS

Clothes moths are major pests of fabric and other items made of natural fibers. Clothes moth larvae commonly feed on wool, feathers, fur, hair, upholstered furniture, animal and fish meals, milk powders, and most animal products, such as bristles, dried hair and leather. Larvae will also infest or feed on lint, dust, paper and materials soiled with oil. Clothes moths can feed on mixtures of natural and synthetic fabrics. However, they cannot feed on materials made of synthetic fibers. In nature, clothes moths have been found infesting pollen, hair, dead insects and dried animal remains.

The most common clothes moths are the webbing clothes moth and the case-making clothes moth. Adult moths do not feed on fabrics, only the caterpillars damage fabrics. Fabrics injured by clothes moths have holes eaten through them by the tiny white caterpillars. Damaged fabrics often have silken cases or silken threads on the surface. Adult moths may be found running over the surface of infested garments or materials. Unlike many other moths, clothes moths are not attracted to light and avoid lighted areas.

Description

Clothes moths (Figure 1) are small yellowish or brownish moths less than 1/2 inch in length. The head and front wings of the webbing clothes moth are golden or yellowish in color; the wings do not have spots. The case-making clothes moth has a dusty, brownish head and front wings with 3 dark spots.

The larvae spin a silken tube or case to protect them from the environment and natural enemies. The tubes created by the webbing clothes moth are attached to and are often located in dark protected areas such as seams or hems. Therefore, the larvae of the webbing clothes moth is stationary and feeds in one area. The case made by the case-making clothes moth is not attached to the fabric. The larvae drag the case along and are relatively mobile.

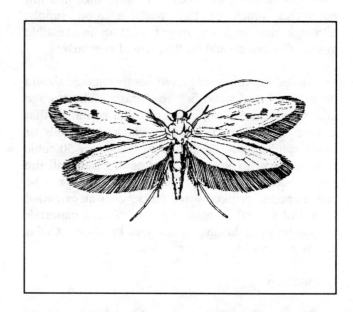

Figure 1. Case-making clothes moth adult.

Life Cycle

The life cycle of the clothes moth can last from 2 months to 2 1/2 years. The adults lay eggs on products that the larvae will consume. Each female moth can lay from 100-150 eggs, which hatch in about 5 days. The small white caterpillars vary in size from 1/16 inch newly hatched to 1/3 inch fully grown. The larval stage varies greatly according to conditions and food supply. The larvae live in cases that are enlarged as they grow. When the larvae pupate, the case is transformed into a tough cocoon. The adult moth emerges in 1-4 weeks.

Control

Prevention and Inspection

The most important method of clothes moth control is good housekeeping. All susceptible articles should be brushed and cleaned periodically, especially items that will be stored for any length of time. Sweep or vacuum regularly to remove woolen lint or hair from floors, shelves and drawers. Also inspect areas for the presence of clothes moths, such as attics, ventilation ducts, and other areas where insects and dust accumulate. Clothing bags, cedar closets and cedar chests only provide protection when stored materials are free from infestation.

Chemical Control

Some woolen fabrics and carpets are mothproofed by the manufacturer; however, less than 20% of susceptible products are mothproofed today. Spot treatment with insecticides may be necessary when clothes moths become established in the home. Apply sprays according to label directions and do not apply directly to clothing. Sprays are effective when properly applied to surfaces as spot treatments. Sprays should be directed to all known or suspected breeding places. Clothing should be removed from closets and drawers before spraying interior surfaces. Space spray aerosols will kill flying moths, but provide no residual protection. Dusts may be used around felts on pianos, under rugs, or on fabrics which may be stained by sprays. To insure protection, treatments may be applied two times per year. Tight closets, trunks, or chests can be mothproofed by application of moth balls at the rate of 1 pound per 50 cubic feet of space.

PLASTER BAGWORMS

Plaster bagworms are similar in appearance and closely related to clothes moths. The larvae of bagworms live in a flattened, gray, watermelon seed-shaped case about 1/2 inch long. The case is constructed of silken fiber and sand particles, lint, paint fragments, and other debris attached. The case has a slit-like opening at each end, and the larva is able to move around and feed from either end.

Plaster bagworms are easily seen on light-colored walls. Close examination of the house may reveal bagworms attached to the underside of chairs, bookcases, and other furniture. They are often found along the edge of rugs, near baseboards, or on the lower edges of walls. Bagworms are quite common in garages and underneath buildings. The larvae mainly feed on spider webs and webs; however, they will also feed on fabrics made of natural fiber.

Control of plaster bagworms is similar to control of clothes moths. Good housekeeping is important, especially the removal of spider webs. Sweep down and remove any spider webs and bagworm cases.

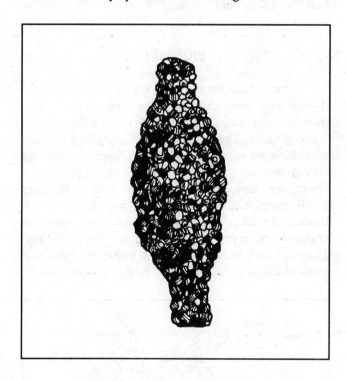

Figure 2. Plaster bagworm.

Subterranean Termites

The three principal types of termites in Florida are subterranean (nest in the soil), dampwood (infest dampwood), and drywood termites (infest dry wood). Subterranean termites are the most destructive and frequently encountered kind of termite found throughout the state. Although they nest in soil, subterranean termites can attack structures by building tubes that connect their nest to wood in structures.

BIOLOGY

Subterranean termites are social insects that live in colonies consisting of many individuals. The colonies are composed of workers, soldiers and reproductives. The workers, which are about 1/8 inch long, have no wings, are white to cream colored and very numerous. Soldiers defend the colony against insects, like ants, that can attack the colony. Soldiers are wingless and white in color with large brown heads and mandibles (jaws). King and queen termites perform the reproductive functions of the colony. They are dark brown to black in color and have two pairs of wings about twice the length of their body.

Subterranean termites feed on wood or other items that contain cellulose, such as paper, fiberboard, and some fabrics derived from cotton or plant fibers. Termites have protozoa in their digestive tracts that can convert cellulose into usable food.

Subterranean termites nest in the soil to obtain moisture, but they also nest in wood that is often wet. They easily attack any wood in contact with the ground. If the wood does not contact the soil, they can build mud tunnels or tubes to reach wood several feet above the ground. These tunnels can extend for 50-60 feet to reach wood and often enter a structure through expansion joints in concrete slabs or where utilities enter the house.

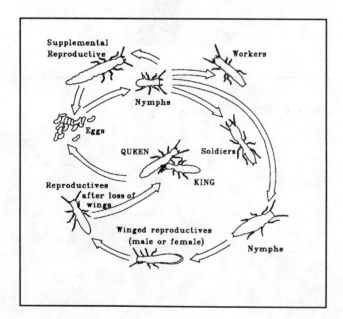

Figure 1. Termite Life Cycle.

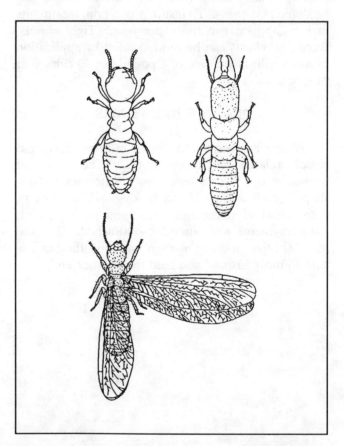

Figure 2. Castes of termite from top left, right, and bottom. Worker, Soldier and Winged Reproductive.

DETECTION OF TERMITES

Termites remain hidden within wood and are often difficult to detect. However, subterranean termites may be detected by the presence of winged reproductives, mud tubes, and wood damage.

Winged Reproductives

Winged reproductives emerge from colonies in great numbers usually in the spring and during the daylight hours. Usually termites are first noticed by the presence of winged reproductives. Mating occurs during these flights, and males and females form new colonies. Winged termites can be distinguished from flying ants by their thick-waist, straight antennae and wings of equal size.

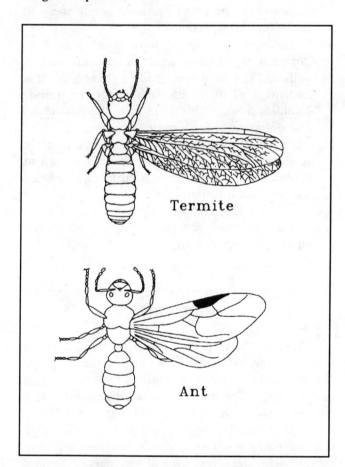

Figure 3. Subterranean termite and winged ant reproductive. Termites have thick-waist, straight antennae and wings of equal size.

Winged termites in a house are an indication of probable infestation. Termite wings break off shortly after their flight, and even though the actual swarming is not observed, the presence of discarded wings indicate that a colony is nearby. Because termites are attracted to light, their broken-off wings are often near doors or windows where the termites have been attracted to the light.

Winged termites emerging from the ground out-of-doors near the house does not necessarily mean the house is infested, but it is a good reason to check further. Termites in the wood of homes or other buildings usually come from colonies already established in the soil.

Peak swarming periods for subterranean termites are from January through May in Florida. They may also swarm to a lesser extent during the other months.

Mud Tubes

Subterranean termites build earthen, shelter tubes to protect them from low humidity and predation. These tubes are usually 1/4 to 1 inch wide. Houses should be inspected at least once a year for evidence of tubes. If the house has a crawl space, the inside and outside of foundations should be inspected for tubes. If the house has a concrete slab floor, cracks in concrete floors and places where pipes and utilities go through the slab should be closely examined. Cracks in concrete foundations and open voids in concrete block foundations are also hidden avenues of entry.

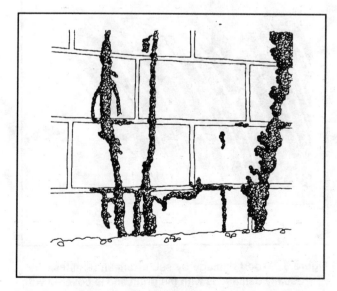

Figure 4. Mud tubes connect the colony in soil with wood in the structure.

Wood Damage

Wood damaged by subterranean termites is often not noticed because the exterior surface usually must be removed to see the damage. However, galleries can be detected by tapping the wood every few inches with the handle of a screwdriver. Damaged wood sounds hollow, and the screwdriver may even break through into the galleries.

Subterranean termite feeding follows the grain of the wood and only the soft springwood is attacked. Unlike drywood termites or other wood-boring insects, subterranean termites do not push wood particles or pellets (fecal material) to the outside, but rather use it in the construction of their tunnels. This debris, along with sand and soil particles, is used as a form of plaster.

Identification

Subterranean and drywood termites require completely different control methods; therefore, the termites must be correctly identified. Soldiers, winged specimens or wings can be identified at your county extension office. Workers and immatures are virtually impossible to identify. If you decide that the services of an experienced pest control operator are needed, contact at least two or three reputable firms in your area for inspections and estimates for treatment.

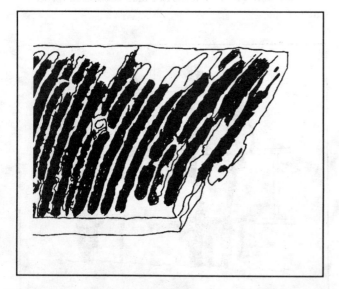

Figure 5. Wood damage by subterranean termites.
Usually damage is with the grain and is covered with muddy material.

SUBTERRANEAN TERMITE PREVENTION AND CONTROL

The best control of subterranean termites is prevention. The best time to provide protection against termites is during the planning and construction of a building. Prevention should include:

1. Removal of all stumps, roots, wood, and similar materials from the building site before construction is begun.
2. Removal of all form boards and grade stakes used in construction.
3. There should be no contact between the building woodwork and the soil or fill. Exterior woodwork should be located a minimum of 6 inches above ground and beams in crawl spaces at least 18 inches above ground to provide ample space to make future inspections.
4. Ventilation openings in foundations should be designed to prevent dead air pockets and of sufficient size to assure frequent changes of air — at least 2 sq. ft. to 25 running feet of outside foundation wall. This helps keep the ground dry and unfavorable for termites.
5. Thorough annual inspections should be conducted to discover evidence of termite activity such as shelter tubes on foundation surfaces, discarded wings or adult termites.
6. Any wood that contacts the soil, such as fence posts, poles and general foundation structures, should be commercially pressure treated.

PRECONSTRUCTION TREATMENT OF STRUCTURES

Control subterranean termites by preventing the termite colony in the soil from entering the structure. It is impossible to build structures so termites cannot cause damage. Therefore, a thorough preconstruction treatment should be applied to protect the structure for 5-20 years. Houses treated prior to 1988 with chlorinated hydrocarbons, should be protected from subterranean termites for 30-40 years. Several insecticides have proven satisfactory for making effective barriers when properly applied.

PRECONSTRUCTION TREATMENT OF FOUNDATION WALLS AND PIERS

After the footings are poured and the foundation walls and/or piers have been constructed, apply the insecticide to a trench in the soil about 6-12 inches wide and 4-6 inches deep adjacent to the foundation. The insecticide must be applied to both the inside and outside of the foundation and also around piers, chimney bases, pipes, conduits and any other structures in contact with the soil. The trench should be as deep as the top of the footing. The insecticide should be mixed with water as recommended on the pesticide label and applied at the rate of 2 gallons per 5 linear feet of trench. The insecticide should be mixed with the soil as it is being replaced.

PRECONSTRUCTION TREATMENT OF CONCRETE SLABS

The most common type of construction in Florida is concrete slab resting on the soil. Often the slabs crack or shrink away from the foundation wall allowing termites to infest the wood above.

The soil underneath and around the concrete slab should be treated with insecticide before the concrete slab is poured. The chemical should be applied after all the subslab fill and reinforcement rods are in place. Apply diluted spray to the fill at the rate of 1 gal. per 10 sq. ft. Along both sides of foundation walls and interior foundation walls and plumbing (critical areas), apply diluted insecticide at the rate of 2 gal. per 5 linear feet. Treat all hollow masonry units of foundations with 1 gal. of diluted spray per 5 linear feet. Apply the insecticide to reach the footing.

POSTCONSTRUCTION TREATMENT OF STRUCTURES

Crawl Space Treatment

Dig narrow trenches along both the inside and outside of foundation walls and around piers and chimney bases, and apply diluted spray as described above. Also be sure to trench and treat around sewer pipes, conduits and all other structural members in contact with the soil. Apply the insecticide to the trenches. The insecticide must be applied to both the inside and outside of the foundation and also around piers, chimney bases, pipes, conduits and any other structures in contact with the soil. The trench should be as deep as the top of the footing. Mix the insecticide with water as recommended on the pesticide label. Apply the diluted spray at the rate of 2 gal. per 5 linear feet of trench. Mix the insecticide with the soil as it is being replaced.

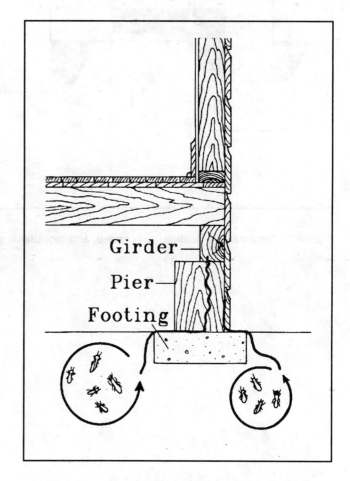

Figure 6. Infestation of crawl space construction.

Concrete Slab Construction

It is possible to trench around the outside of a slab after it has been poured, as described above, but this alone usually will not give satisfactory control because the termite colony may be entering the structure from the soil under the slab.

Homeowners are not equipped to treat under slabs after the slab foundation is completed. A professional pest control operator usually is needed to do subslab chemical injections.

128

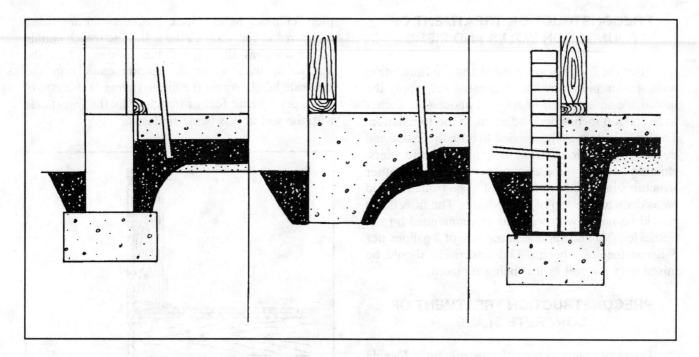

Figure 7. Treatment of floating, suspended, and monolithic concrete slabs.

Formosan Subterranean Termite

The Formosan subterranean termite, *Coptotermes formosanus* Shiraki, is considered one of the most destructive and aggressive species of termites in the world. In 1980 it was found causing damage to a Hallandale condominium, in 1982 a house in Orlando, 1984 a house in Gulf Breeze, and in 1991 a condominium in Tampa. On a 1982-3 survey conducted by Thompson (1985), the Formosan subterranean termite was collected over approximately 60 sq. miles covering Broward and Dade counties. Of the 40 condominium lots surveyed, 17 had active infestations. Data collected by Su and Scheffrahn indicate this figure has grown substantially. The Formosan subterranean termite is known to damage buildings, living trees, utility poles and railroad ties.

HISTORY AND SPREAD

The Formosan subterranean termite is native to China and has been introduced into Japan, Guam, Sri Lanka, South Africa, Hawaii and the continental United States. It was first discovered at a Houston, Texas shipyard in 1965. In 1966, well established colonies of the Formosan subterranean termite were discovered in New Orleans and Lake Charles, Louisiana and Houston and Galveston, Texas. In 1967, the species was found in Charleston, South Carolina. Well-established colonies were located in Florida in 1980, 1982 and 1984. The Formosan subterranean termite is extremely destructive, where it has become established.

Most introductions and spread of the Formosan subterranean termite throughout the world have probably been from ships. Once introduced, swarming is the termite's natural method of spread, but the Formosan termite is a weak flier and does not spread rapidly by itself. The winged reproductives are attracted to lights. The movement of infested soil or material such as lumber, wooden crates or other wood products is another important method of spread.

In the United States, the Formosan subterranean termite generally has been confined to the southeast

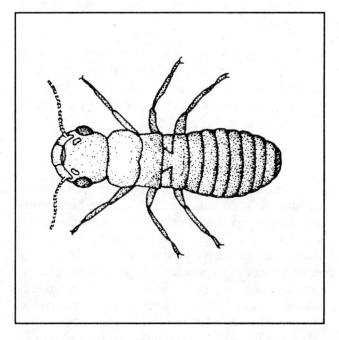

Figure 1. Formosan termite.

at about 32.5° N latitude. This latitude coincides with the warmer temperatures usually associated with this termite. However, the belief that severe winter temperatures limit the spread of these termites may not hold true due to the widespread use of central heating in the U.S. Central heating provides a warm environment, conducive to the survival for termites during winter.

BIOLOGY AND HABITS

Swarming

Termites initiate a new colony by sending out winged reproductives (alates) from an established colony. Major swarms of the Formosan subterranean termite begin in May or June until about July or August on humid, still evenings between dusk and midnight. Because alates are attracted to lights, large numbers can be seen around light sources during a swarm. After a short flight, alates drop to the ground, shed their wings and pair off. If they successfully find a small crevice containing moist wood, the pair forms a chamber in which the eggs are laid. It usually takes 3-5 years to develop a mature colony. A Formosan subterranean termite queen lays approximately 2000 eggs per day. A mature colony averages between 2-8.5 million termites (Lai 1977, Su 1991) and its foraging territory may range up to 300 feet from the nest (King and Spink 1969, Li et al. 1976, Lai 1977).

Swarming generally starts at dusk; however, the Florida termite, *Prorhinotermes* simplex, and some drywood termite species are also night swarmers. Native subterranean termites generally swarm during the day.

Food

Like other termites, Formosan subterranean termites feed on cellulose. They have been found attacking 47 species of living plants including citrus, sugar cane, avocado, wild cherry, cherry laurel, ligustrum, hackberry, cedar, willow, tallow, wax myrtle, sweet gum, mimosa, cypress, red bud, Chinese elm and white oak. Surveys in New Orleans, Louisiana indicated large numbers of residential and city-owned trees were infested. Formosan termites attacks the bases of poles, old tree stumps or other wood in contact with the soil. They have been known to construct galleries to the upper stories of buildings to feed on wood.

The Formosan termite has been known to attack non-cellulose material such as thin sheets of soft metal (lead or copper), asphalt, plaster, mortar, creosote, rubber and plastic in search of food and moisture. Their highly publicized ability to chew through concrete is a fallacy, However, the Formosan subterranean termite is uncanny in finding small cracks in concrete which they use are foraging routes.

A single Formosan subterranean termite does not consume more wood than does the native subterranean termites (Su and LaFage 1984a and b). The rapidity of damage is related to the termite's high reproductive capacity and large colony size.

Nests

Formosan termite nests are made of carton which consists of chewed wood, saliva and excrement. Nests can be constructed in the ground or aerially.

Formosan subterranean termites can produce massive carton nests. The native subterranean termite also produces carton when nesting, but it is small when compared with the Formosan subterranean termite nest. Auxiliary nests are often constructed in the walls of buildings or in the food source. When a nest or gallery is disturbed, the Formosan termite soldiers are usually present in larger numbers and are more aggressive than the native termites.

Damage

Most subterranean termites feed along the grain of the wood, eating the spring wood and leaving the summer wood. The Formosan termite feeds on both and forms a hollow. If the hollow is large (tree or timber), it is then filled with carton material to form a nest. Formosan termites, like our native subterranean termite, construct tunnels (foraging galleries) from the nest to infest wood. In Hawaii, where unprotected homes were built over large colonies, records show that the Formosan subterranean termite caused major structural damage in 6 months and almost complete destruction in 2 years (Tamashiro 1984).

Moisture Requirements

The Formosan termite, like all subterranean termites, uses the soil for a source of moisture. However, Formosan termite colonies can obtain moisture from plumbing or roofing leaks.

RECOGNITION

Termites are social insects and 3 forms (called castes) are commonly seen: winged reproductives (alates), soldiers and workers. The nests are also characteristic and easily identifiable as Formosan.

Soldiers

The Formosan termite soldier has an oval or egg-shaped head, which is shorter than the heads of native termites. When an infested timber is examined, soldiers are more abundant than with our native termites. The soldiers are more aggressive and will attack anything used to probe the damage. When disturbed, the soldiers exude a white substance (defensive secretion) from the fontanelle, located on the top, front of the head. In native termites, the fontanelle is small and barely visible.

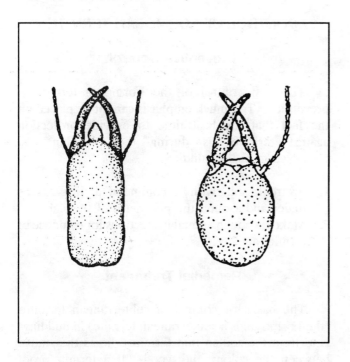

Figure 2. Head capsule of soldier termite.

Winged Reproductives

The winged reproductives of the Formosan termite are yellowish-brown and are 12-15 mm long (0.5-0.6 inches). Native subterranean termites are 10 mm long (0.4 inches). *Reticulitermes virginicus* and *R. flavipes* have black bodies, while *R. hageni* is about the same color as the Formosan termite. The ocelli of the Formosan termite are large, whereas they are small on native termites.

The winged forms look very much like drywood termite reproductives in size and color. It is important to look at the head or wings with a microscope to tell the difference positively. Otherwise, it is necessary to find soldiers or damage.

The wings of the Formosan termite are 10 mm in length and have two heavily pigmented veins near the front edge of the wing. The median vein is present and may or may not be branched. Formosan subterranean termite wings are hairy when compared with the nearly bare wings of the native subterranean termite when observed under low magnification.

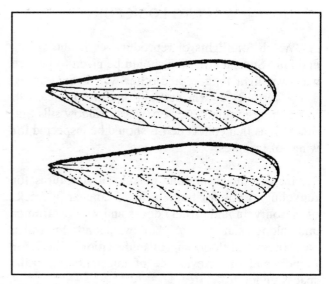

Figure 3. Formosan termite wing venation.

Workers and Nymphs

The workers and nymphs of the Formosan termite are white. They are difficult to distinguish from other termite species.

WHERE TO LOOK FOR FORMOSAN TERMITES

All types of structures, piers, pilings, posts, logs, utility poles and timber in contact with soil should be carefully inspected.

Formosan termites generally invade structures from the ground. They commonly enter though expansion joints, cracks, and utility conduits in slabs, and holes for tub drains. Any wood connecting with the ground is an inviting entrance to this termites. Foraging galleries lined with carton material are then constructed to the food source.

The Formosan subterranean termite does not always require ground connection. If a pair of alates successfully finds adequate food and moisture sources in a building, it can initiate a colony with no ground connection. The flat roofs of high rise buildings are places for the Formosan subterranean termite to initiate aerial infestations if portals of entry are found because they almost always contain water. Survey data indicate that more than 25% of the infestations found in the urban southeastern Florida are caused by aerial colonies.

WHAT TO LOOK FOR

Watch for flights of reproductives in the spring and fall. Special attention should be given to picture windows, light fixtures and other well lighted areas. These termites swarm at night and are attracted to lights in large numbers. Cobwebs, window sills and other areas that collect debris should be inspected for wings of termites.

Inspect the inside and outside of structures for tunneling. Look for tunnels that disappear in cracks of masonry, in and around doors and window frames and along siding. A hollow sound in walls, baseboards and floors suggests infestation. Check for tunnels and other evidence of infestation in crawl spaces under structures.

Tunnels may not be evident in wood or wood products. It is best to probe such samples for evidence of infestation.

INSPECTION OF CONCRETE SLAB CONSTRUCTION

Inspect for evidence of termite activity near any plumbing that goes through the slab. Look for tubes around baseboards. Tap baseboards around walls. Check for wood which is in contact with the soil.

IDENTIFICATION OF SPECIMENS

Collect evidence of infestation and save soldiers or winged reproductives. Examine the specimens closely or send suspect samples to your County Cooperative Extension Office. Preserve all specimens in 70% rubbing alcohol. Submit at least ten (10) specimens if available.

PROCEDURES FOR IDENTIFICATION

1. To identify the winged reproductive, make sure the wing is completely flattened. Often wings have a tendency to curl along the front edge, and the wing may appear to have 1 or 2 veins rather than 2 or 3.
2. Place the wing on a white background to obtain a true wing color.
3. Observe type of damage.
4. Observe how wet or dry the wood is near the area of termite damage.

5. Note the time of year and time of day swarming took place.

CONTROL OF FORMOSAN TERMITE

Preventive Control

The best control of the Formosan termite is prevention. Too much emphasis cannot be placed on the fact that the best time to provide protection against termites is during the planning and construction of the building.

1. Pretreat soil with recommended insecticides under and around the perimeter of the slab.
2. Make thorough annual inspections for evidence of termite activity.

Remedial Treatment

The basis for control of subterranean termites (the Formosan is a subterranean termite) in buildings is to treat the soil so that termites die as they move between the colony and wood. If a termite-proof chemical barrier between the soil and wood is maintained, the termites in the house cannot gain access to the soil to get moisture. If there is no moisture in the house, the termites will die. Also termites in the soil cannot gain access to the wood in the house.

Spot treatment of sections of a building may not prevent the termite from gaining access elsewhere. The Formosan termite constructs galleries that may cover up to 1 acre of land, and it can easily find untreated areas to enter the building. Every effort should be made to complete the chemical barrier under the building to prevent future entry. Control is possible if the pest control operator and the public are aware that greater care is required when treating infestations. The Formosan termite takes advantage of a pesticide applicator's mistakes and is likely to require retreatment more often than native termites.

Buildings should be thoroughly inspected to discover and eliminate all sources of moisture. The Formosan termite builds auxiliary nests in the walls of buildings and is able to survive for months without soil contact. Therefore nests should be located and a chemical treatment applied to destroy the nests.

Home owners should consult a reputable, experienced, certified pest control operator rather than attempt home remedies to control this serious pest.

Fumigation

Generally fumigation is not recommended to control Formosan termites. It may be necessary to apply a fumigant gas to kill auxiliary nests in special cases where the soil has been properly treated and the infestation continues from an unknown moisture source. If a house is fumigated and the soil is not treated, there is nothing to prevent reentry of termites from the soil.

Treatment of Infested Trees

Surface applications of chemicals will not control Formosan termites, which usually hollow out the center of trees. Formosan termites can be controlled in living trees by drilling holes above the soil line and injecting the chemical into the void created by the termites.

How to Choose a Proper Pest Control Company

The following are suggestions made by the Florida Pest Control Association:

-The pest control company should make a complete inspection of the entire building to determine the origin and the extent of infestation. This is extremely important, as without a thorough inspection of infestation, proper treatment cannot be recommended.

-A written report as to the extent of the infestation and probably origin (aerial or ground) should be given.

-Ask if the contract makes any distinction between the Formosan subterranean termite and the native subterranean termite.

-Any company whose contract makes a distinction between these two types of termites probably has enough experience because it realizes the need for this separation.

-Ask for references of previous Formosan subterranean termite work.

-Do not be pressured by a company thatleads you to believe you have a serious situation, which should be treated immediately. Ask if they object to a second opinion concerning the method of treatment or extent of damage.

-Lastly, ask the company to contact the University of Florida, IFAS Extension Office, to record pertinent information about theFormosan subterranean termite infestation.

-The key to any decision is confidence in the company you chose. Pay attention to value and service you expect for the price you pay.

REFERENCES CITED

King, E.G. and W.T. Spink. 1969. Foraging galleries of the Formosan subterranean termite, *Coptotermes formosanus*, in Louisiana. Ann. Entomol. Soc. Am. 62:537-542.

Lai, P.Y. 1977. Biology and ecology of the Formosan subterranean termite, *Coptotermes formosanus*, and its susceptibility to the entomomogenous fungi, *Beaveria bassiana* and *Metarrhizium anisopliae*. Ph.D. dissertation, University of Hawaii, Honolulu.

Li, T., K.H. He, D.X. Gao, and Y. Chao. 1976. A preliminary study of the foraging behavior of the termite, *Coptotermes formosanus* (Shiraki) by labeling with Iodine-131. Acta Entomol. Sinica. 19:32-38.

Su, N.-Y., and J.P. LaFage. 1984a. Comparison of laboratory methods for estimating wood-consumption rates by *Coptotermes formosanus* (Isoptera: Rhinotermitidae). Ann. Entomol. Soc. Am. 77:125-129.

Su, N.-Y., and J.P. LaFage. 1984a. Differences in survival and feeding activity among colonies of the Formosan subterranean termite (Isoptera: Rhinotermitidae). A. Ang. Ent. 97:134-138.

134

Su, N.-Y., P.M. Ban, and R.H. Scheffrahn. 1991. Suppression of foraging populations of the Formosan subterranean termite (Isoptera: Rhinotermitidae) by field applications of a slow-acting toxicant bait. J. Econ. Entomol. 84(5):1525-1531.

Tamashiro, M. 1984. The Formosan subterranean termite. Whitmire Institute of Technology and Advanced Pest Management. 4pp.

Thompson, C.R. 1985. Detection and distribution of Formosan termite (Isoptera: Rhinotermitidae) in southeastern Florida. J. Econ. Entomol. 78:528-530.

Non-Subterranean Termites

DRYWOOD TERMITES

Drywood (non-subterranean) termites as well as subterranean termites occur in Florida. The former infest dry wood and do not require contact with the soil.

Drywood termites are social insects that live in colonies. The colonies are composed of kings, queens and soldiers. There is no worker caste as in subterranean colonies. The work is performed by immature termites before they become adults. King and queen termites perform the reproductive functions of the colony. They are light to dark brown and 1/3 to 1 inch in length. Soldiers guard the colony against invaders such as ants. They are white and wingless with large brownish heads and jaws. The nymphs (immatures), which are the most numerous caste, are white and wingless. The soldiers and immatures remain inside the wood at all times.

Subterranean termites occur throughout the state while drywood termites are more common along coastal areas although they are also found inland. Termite food consists of cellulose obtained from wood. Protozoa in the termites' digestive tracts convert the cellulose into usable food. Infestations of drywood termites may be found in almost any product containing cellulose. This insect is most commonly found infesting woodwork in buildings and furniture.

Non-subterranean termites remain hidden within the wood or other material on which they feed, so that those actually feeding are seldom seen. Galleries or tunnels in the wood made by drywood termites cut across the grain of the wood and destroy both soft spring wood and the harder summer growth. Galleries made by the subterranean species follow the grain of the wood and attack only the soft spring wood.

Signs of Infestation

There are several signs of drywood termite infestations. At certain times of the year during

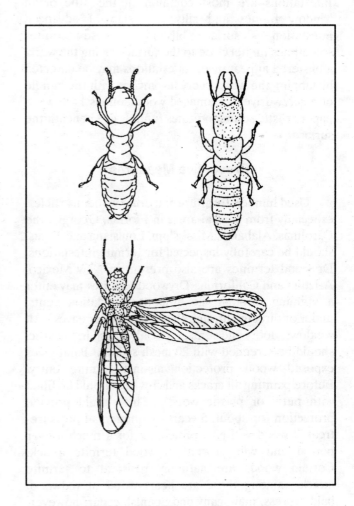

Figure 1. Drywood Termite.

daylight hours, king and queen termites emerge from the colonies. The purpose of these flights is to establish new colonies. Peak swarming periods are from January through May in Florida although they may occur to a lesser extent during other months. Winged termites can be distinguished from winged ants because termites have a thick waist whereas ants have a distinctly thin or wasp-like waist. The appearance of winged termites in the home is an indication of probable infestation; however, they may come in from outside. Wings break off shortly after the termites swarm and because they are attracted to light at this stage, their discarded wings are often found on window sills.

Drywood termites unlike subterraneans excrete pellets of partly digested wood. These pellets are straw-colored to reddish-brown and about the size of sand. The pellets are pushed from the galleries and found on surfaces beneath the infested wood. Infestations are most common in the attic or in window frames and sills. A sign of advanced infestation is surface blisters. These termites sometimes tunnel close to the surface giving the wood a blistered appearance. Infestations may be detected by tapping the wood every few inches with the handle of a screwdriver. Damaged wood sounds hollow - a papery rustle sound indicates tunnels just beneath the surface.

Preventive Measures

Used lumber, furniture and other wooden articles, especially from coastal areas in Florida, Georgia, the Carolinas, Alabama, Mississippi, Louisiana and Texas, should be carefully inspected for termite infestations. Drywood termites are also present in New Mexico, Arizona and California. Drywood termites may enter a building through the attic or foundation vents, under or directly through shingles or under eaves. All windows, doors and vents, especially those in the attic, should be screened with 20-mesh screen. Paint gives exposed wood protection against termite entry. Before painting all cracks and crevices should be filled with putty or plastic wood. This should provide protection for about, 5 years. Commercial pressure-treated wood will give protection for a much longer period and will prevent drywood termite attack. Certain woods are naturally resistant to termite attacks; among these are heart wood of redwood, bald-cypress, mahogany and Spanish cedar; however, these woods will become susceptible after several years after weathering.

Control

Drywood and subterranean termites require completely different control methods; therefore the termites must be correctly identified. If in doubt, take several soldiers, winged specimens or broken-off wings to your County Extension Office. Immatures (workers) are virtually impossible to identify.

If detected in the early stages and damage is localized, a drywood termite colony may be controlled by removing and replacing the damaged wood or by the application of an insecticide. It is very important to carefully inspect all woodwork of the entire building, especially attics, baseboards, window sills, floor joists and furniture for termite pellets and/or damaged wood.

If the infestation is too extensive and advanced for local treatment, it will be necessary to tent and fumigate the entire building. Although this method is very expensive and leaves no residual protection, it is usually the only alternative when many termite colonies are present. Fumigation can only be performed by a licensed pest control operator. Contact several firms for inspections and estimates for treatment.

Fumigant, gases used for structural fumigation are 100% effective when adequate concentrations are maintained within the structure. Proper fumigation will not damage household products.

FLORIDA DAMPWOOD TERMITE

Colonies are made up of reproductives, soldiers, and workers (nymphs). Reproductives are 1/2 to 1 inch long, light to dark brown and have 2 pair of wings of equal length with 2 pigmented veins and no median vein. Soldiers are 1/2 inch long, wingless, dark brown in color, mandibles without teeth and head wider than pronotum, oval-shaped with sides convex. Workers are about 1/2 inch long, wingless and white to cream in color. The Florida dampwood termite nests in damp wood.

Signs of Infestation

Signs of Florida dampwood termite infestations include small six-sided fecal pellets in dry situations, muddy excrement in wet situations and hollow-sounding wood with galleries cutting across the grain.

Control

For non-chemical control, construction should be designed to eliminate moisture and water leaks, and wood near soil should be treated with a preservative. Replace infested or damaged wood.

To chemically control, treat wood with termiticides registered for wood treatment.

SOUTHERN DAMPWOOD TERMITE

Colonies are made up of reproductives, soldiers, and workers (nymphs). Reproductives are 1/2 to 1 inch long, light to dark brown, and have 2 pair of wings of equal length with 3 or more pigmented veins and median vein with numerous cross veins. Soldiers are about 1 inch long, wingless, brown in color, mandibles with teeth, head elongate, and antennae with 3rd segment same size as 2nd and 4th. Workers are about 1/2 inch long, wingless and white to cream in color.

The Southern dampwood termite nest in dampwood and attack both soft springwood and harder summer growth.

Signs of Infestation

Signs of Southern dampwood termite infestation include small, hard dry fecal pellets pushed to outside of wood work and hollow-sounding wood with galleries cutting across the grain.

Control

For non-chemical control, construction should be designed to eliminate moisture and water leaks, and wood near soil should be treated with a preservative. Remove and replace infested or damaged wood.

To chemically control, treat wood with termiticides registered for wood treatment.

138

Powder Post Beetles and Other Wood-Infesting Insects

ANOBIID POWDER-POST BEETLE

The anobiid powder-post beetle is 1/16 to 1/4 inch long, reddish brown or grayish brown to dark brown. Its body is cylindrical, elongated and covered with fine hair. Its head is covered by pronotum from top view; the last 3 segments of antennae are usually lengthened and broadened. The larva are white, "C" shaped with rows of small spines on dorsal side. This beetle's life cycle is 1 to 10 years. Anobiids infest seasoned softwood and the sapwood of seasoned hardwoods.

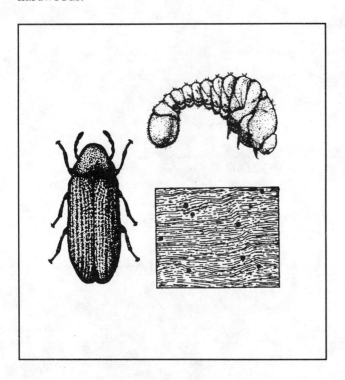

Figure 1. Deathwatch Beetle (Anobiidae).

Signs of Infestation

Emergence holes from anobiids are 1/16 to 1/8 inch in diameter, round in shape. Cigar-shaped frass found loosely in tunnels and in small mounds outside of emergence holes.

Control

If infestation is localized, treat wood with a residual spray; if well advanced, fumigate the structure.

Non-chemical control may be obtained by sanding, filling cracks then painting, varnishing or waxing. Replace wood if damage is extensive.

BOSTRICHID POWDER-POST BEETLE

Bostrichid powder-post beetles are 1/8 to 3/4 inch long, reddish brown to black in color. Their bodies areelongated and cylindrical with roughened thorax. Heads are deflexed and concealed by pronotum from top view; antennae are short with 3 or 4 enlarged sawtoothed terminal segments. The larva is white, "C" shaped with no spines on body. The life cycle is 1 year. Bostrichids infest seasoned softwood and hardwood; especially unfinished floors, window sills, furniture, etc.

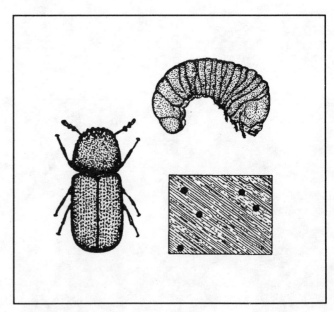

Figure 2. Bamboo Borer (Bostrichidae).

Signs of Infestation

The round emergence hole is 1/8 to 3/16 inch in diameter. Sawdust-like frass sticks together and is found tightly packed in galleries but not in entrance holes.

Control

If infestation is localized, treat wood with a residual spray; if well advanced, fumigate the structure. Replace wood if damage is extensive.

Non-chemical control may be obtained by sanding, filling cracks, then painting, varnishing or waxing. Replace wood if damage is extensive.

LYCTID POWDER-POST BEETLES

Lyctid powder-post beetles are about 1/4 inch long, brown in color, body elongated and slightly flattened, prominent head not covered by pronotum, antennae with a 2-segmented terminal club. The larvae are white, "C" shaped, with 8th abdominal spiracle enlarged. This beetle's life cycle is 6 months to 4 years. Lyctids infest the sapwoods of hardwoods; mainly ash, hickory, oak, maple and mahogany.

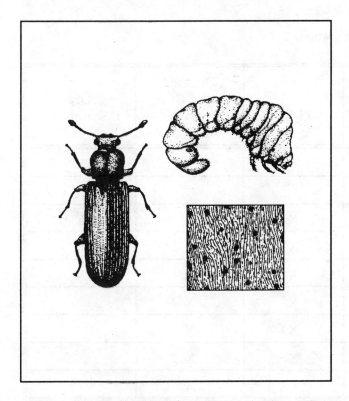

Figure 3. Powder-Post Beetle (Lyctid).

Signs of Infestation

The round emergence hole is 1/16 to 1/8 inch in diameter. A fine powder-like dust is found loosely packed in tunnels and in small piles outside of the hole.

Control

If infestation is localized, treat wood with a residual spray; if well advanced, fumigate the structure.

Non-chemical control may be obtained by sanding, filling cracks, then painting, varnishing or waxing. Replace wood if damage is extensive.

OLD HOUSE BORER

The larvae of a long-horned beetle (old house borer) sometimes infests seasoned softwoods in Florida. The eggs are laid in crevices of the bark of cut logs and the larvae require 3 to 5 years or more to mature. The larvae are white, segmented and have an enlarged, brownish head. They bore through the sapwood making irregular galleries. During quiet times, their feeding may be heard as a clicking or rasping sound. The beetles emerge through an oval hole about 1/4 inch in diameter. The adult beetle is grayish-black, 1/2 to 3/4 inch in length with several white markings on the wing covers and long antennae. Re-infestation in buildings is not common.

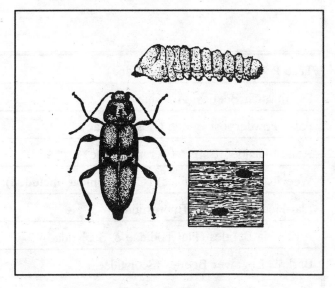

Figure 4. Old House Borer (Cerambycidae).

140

Control

Localized infestations can usually be controlled as outlined for powder-post beetles. Control is very difficult in large timbers, because it is almost impossible to get an insecticide deep enough into the wood. If the infestation is not confined to small areas, fumigation of the entire structure may be necessary by a licensed pest control operator.

CARPENTER BEES

Carpenter bees are 3/4 to 1 inch long and closely resemble bumble bees except that their abdomen is shiny metallic greenish-black. The abdomen of bumble bees is very hairy. These insects sometimes build their nests in solid wood such as weather boarding, railings, supports and trim of buildings. Their nests are in the form of tunnels 3 to 6 inches deep in the wood. The entrance hole is about 1/2 inch in diameter. The holes are very clean and appear as though they were made by a drill. Damage to wood is seldom extensive.

Figure 5. Carpenter Bee.

Control

Carpenter bees can be controlled by blowing a small amount of dust into the holes. After a few days, plug the holes with plastic wood, putty or a similar material.

Table 1. Classification of families of beetles which attack wood.

	Living Wood	Diseased and/or Dying Trees and Also Logs	Dry Seasoned Wood
True Powderpost Beetle (Lyctidae)			X
Deathwatch Beetles (Anobiidae)			X
False Powderpost Beetle (Bostrichidae)	X	X	X
Round-Headed Borers (Cerambycidae)	X	X	X
Weevils (Several Families) (Curculionidae Included)	X	X	X
Flat-Headed Borers (Buprestidae)	X	X	
Ambrosia Beetles (Platypodidae & Scolytidae)	X	X	
Bark & Engraver Beetles (Scolytidae)	X	X	
Click Beetles (Elateridae)		X	

Occasional Invaders

Pillbugs, Centipedes, Millipedes And Earwigs

SOWBUGS AND PILLBUGS

Sowbugs and pillbugs are common crustacea, belonging to a group of animals called Isopods and are found throughout Florida. They are wingless, oval or slightly elongated arthropods about half an inch in length and slate-gray in color with body segments resembling armored plates.

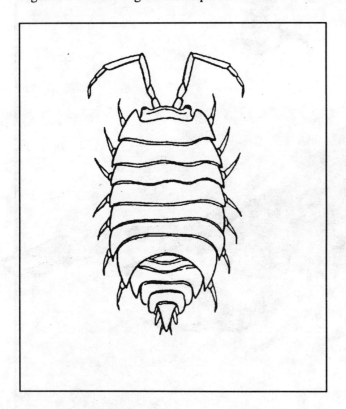

Figure 1. Sowbug.

Both pillbugs and sowbugs feed primarily on decaying organic matter although occasionally they may damage the roots of green plants. Their normal habitat is outdoors, but they occasionally wander inside where they do no damage.

Sowbugs are often called woodlice and possess two tail-like appendages, seven pairs of legs, and well-developed eyes. They are incapable of rolling into a tight ball.

Pillbugs or "rolly-pollies" lack the tail-like appendages and can roll into a tight ball.

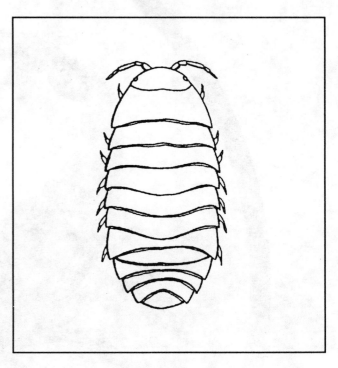

Figure 2. Pillbug.

Habits and Biology

The habits, biology and control of sowbugs and pillbugs are similar. Both animals are slow-moving, crawling arthropods. They require high moisture and are most active at night. When resting during the day, they may be found under trash, rocks, boards, under decaying vegetation, or just beneath the soil surface. A heavy infestation indoors usually indicates a large population outdoors. Mulches, grass clippings and leaf litter often provide the decaying organic matter they need to survive.

Breeding can occur throughout the year in Florida. The female carries the eggs in a brood pouch on the underside of her body. Often there are 7 to 200 eggs per brood. The eggs hatch in 3 to 7 weeks and the young remain in the pouch another 6 to 7 weeks. Once the young leave the pouch, they never return. Some species produce only 1 brood per year, but others may produce 2 or more. Individuals may live up to 3 years.

Control

Sowbugs and pillbugs cause no damage inside the home. Simple mechanical control such as a broom and dustpan or a fly swatter may be adequate. If they become a serious nuisance, elimination of hiding places, food material and moisture sources will reduce the infestation. Source reduction outdoors will help considerably. Piles of leaves, grass clippings and fallen fruit should be removed. Boxes or boards and other debris should be stored off the ground to eliminate a moist shelter.

Pesticide treatment may kill pillbugs and sowbugs which wander inside. Recommended chemicals are 0.5 percent diazinon or chlorpyrifos. Complete control is difficult to achieve and treatments may not last more than one month.

Usually it is necessary to control sowbugs outside. Treatments should be made to and near foundation walls, around steps or damp areas surrounding the structure. Cracks between sidewalks and the foundation require thorough treatment. Recommended chemicals for outdoor treatment are sprays containing 1 percent carbaryl or 0.5 percent diazinon. Granules or dusts containing 2 percent diazinon or 5 percent carbaryl are also useful for treating around foundations and crawl spaces.

CENTIPEDES AND MILLIPEDES

Centipedes and millipedes are commonly seen in yards and occasionally enter homes. Neither centipedes nor millipedes damage furnishings, home or food. Their only importance is that of annoying or frightening individuals.

Centipedes are many-legged animals and belong to a group of animals called Chilopods. They are usually brownish, flattened animals with many body segments. Most of the body segments have one pair of legs. Centipedes are fast runners and may vary in length from one to six inches. They have one pair of antennae or "feelers" that are easily seen.

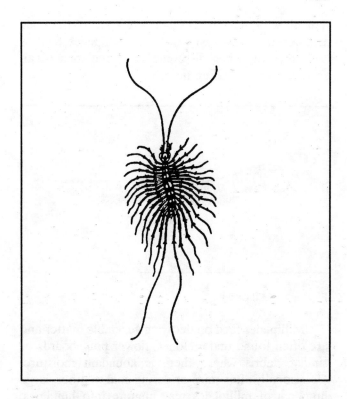

Figure 3. Centipede.

Centipedes have poorly developed eyes and are most active at night. They are active predators and feed mainly on insects and spiders. All centipedes have venom glands to immobilize their prey. The jaws of the smaller local species cannot penetrate human skin; however, the larger species may inflict painful bites.

Centipedes are usually associated with damp, dark places such as under stones, leaf litter, logs, bark or soil crevices. Indoors they may be found in closets and bathrooms where there is high humidity.

Centipedes usually lay 15-55 eggs clustered together in the soil although the eggs of some species are laid singly. The eggs hatch soon after they are deposited. The female will usually guard the eggs and the newly hatched young. Young centipedes closely resemble the adults and require 3 years to mature. Centipedes are rather long-lived and individuals may live up to 6 years.

Millipedes are commonly known as "thousand leggers" and belong to a group of arthropods called Diplopods. Millipedes are worm-like, cylindrical animals with many body segments. Most of the body segments bear two pairs of legs. Millipedes tend to coil up tightly when disturbed and some species can secrete a foul smelling fluid.

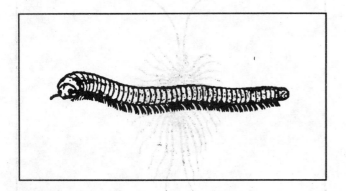

Figure 4. Millipede.

Millipedes feed on decaying vegetable matter and are often found under stones, flower pots, boards or similar debris where there is abundant moisture. Occasionally after rains or during cold weather, large numbers of millipedes may migrate into buildings. They can climb foundation walls and enter homes through any small opening. These pests are generally more troublesome in wooded or newly developed areas where decaying vegetation provide excellent food and breeding conditions.

Female millipedes can lay from 20-300 eggs singularly or in clusters in the soil. The eggs hatch in a few weeks, and the young go through 7 to 8 stages before maturing to adults.

Control

Indoor chemical treatment will eliminate only the centipedes or millipedes already inside. Spot treatments of 0.5 percent diazinon or 0.5 percent chlorpyrifos applied to infested areas will aid in control. Removal of individuals with a broom or dustpan will sometimes be sufficient.

A large indoor population usually indicates large numbers of millipedes or centipedes surrounding the structure. Removal of breeding sites and harborages will aid in control. Compost piles and decaying vegetation should be removed from areas close to the home. Outdoor treatments with sprays containing 1 percent carbaryl or 0.5 percent diazinon should help control outdoor populations. Dusts and granules containing 5 percent carbaryl or 2 percent diazinon

may be applied to crawl spaces and around foundation walls.

EARWIGS

Earwigs are beetle-like, short-winged, fast moving insects about one-half to one inch in length. They are usually dark brown and have a pair of pincer-like appendages at the tip of the abdomen. They have chewing-type mouthparts and slow development.

Earwigs are active at night. They usually hide in cracks, crevices, under bark or in similar places during the day. They are usually scavengers in their feeding habits, but occasionally feed on plants.

The name earwig is derived from an old superstition that these insects enter peoples' ears. This idea is entirely unfounded because earwigs are harmless to man. Some species have scent glands from which they can squirt a foul-smelling liquid. This is probably used for protection; however, it makes them very unpleasant when accidentally or purposely mashed.

The striped earwig adults are dark brown with light tan markings. The males are large and robust with stout pincers. The females are somewhat smaller and lighter in color than the males. These earwigs are in areas having sandy or clay soils, and it lives in subterranean burrows or under debris. They are usually found outdoors unless populations are large or other conditions are adverse. They enter structures in search of food, a more suitable environment or just by accidental meanderings.

Because of their nighttime activity, they remain in the soil or under debris during the day. Heavily thatched lawns or mulched flower beds are among their preferred daytime habitat. At night they collect in large numbers around street lights, neon lights, lighted windows or similar locations where they search for food. Favorite foods include armyworms, aphids, mites and scales. They also forage on food scraps or dead insects.

The female lays about 50 tiny eggs in a subterranean burrow. The eggs hatch into nymphs in about 7 days and the nymphs feed on their egg case. The female continues to care for the young, grooming and manipulating them in the burrow throughout the first nymphal stage. The young nymphs are about one-eighth inch long and could be very easily confused with termites.

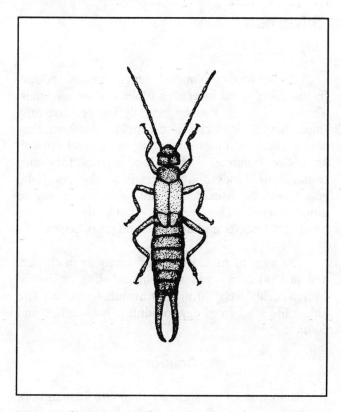

Figure 5. Earwig.

In about seven days, the nymphs molt into the second stage and they are released from the burrow by the female. At this time the female loses her maternal instincts and many times will devour the nymphs before they can hide. During later stages, the nymphs tend to be cannibalistic. After passing through 6 nymphal stages, the earwig becomes an adult, the life cycle egg to adult having taken an average of 56 days.

Control

Indoors

Earwigs are difficult to control with chemicals. Many times the most effective control for an errant earwig or two in the home is purely mechanical. A folded newspaper or fly swatter, a broom and dust pan provide quick and inexpensive control.

If severe infestations occur, insecticide sprays can be applied. These materials should be used only for spot treatment indoors.

Outdoors

Proper scheduling of applications may increase the efficiency of control. Application of one of the recommended materials should be made late in the afternoon or early evening since earwigs are active at night. The material should be applied in a band treatment around the entire perimeter of a structure and the band should be at least 10 feet wide. It may also be necessary to treat the base of mulched shrubbery or flower beds.

Because of the high reproductive potential and habitat of earwigs, it is likely that insecticidal control will have to be repeated at regular intervals.

Sprays may be used around the structure. Dusts or granules are also effective. Glass jars or tin cans baited with fish or cat food can be buried level with the ground line for use as pitfall traps. The earwigs cannot climb the sides of the container and are trapped. The trap can be cleaned periodically and the trapped earwigs destroyed.

146

Booklice And Silverfish

BOOKLICE

Booklice belong to a group of insects collectively called psocids. The psocids are small, soft-bodied insects, most of which are less than 1/8 of an inch long. They are both winged and wingless. Psocids have chewing mouthparts.

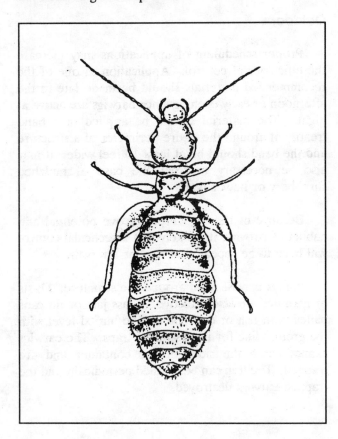

Figure 1. Book louse.

The majority of psocids are outdoor species with well-developed wings. They are most commonly found on bark or foliage of trees and shrubs. These psocids are frequently called "barklice." Most of the species found in buildings are wingless. Because they are often found among books or papers, they are called booklice. The term "lice" is somewhat misleading because none of these insects are parasites and few of them have a louselike appearance.

Psocids feed on molds, fungi, cereals, pollen, fragments of dead insects or other similar materials. They cause little loss by actually eating foodstuffs since they do feed chiefly on mold. At times they may become extremely abundant and spread through an entire building. In such situations they may contaminate foods and materials to the point the goods must be discarded. Damage to books may be more direct. They eat the starch sizing in the bindings of books and along the edges of pages.

The eggs of psocids are laid singly or in clusters and are often covered with silken webs or debris. Most species pass through 6 nymphal stages. The entire life span from egg to adult is between 30 and 60 days.

Control

Reduction of moisture to eliminate formation of mold is a very effective method for controlling booklice. Infested furniture, bedding or other movable furnishings should be thoroughly cleaned and aired. Clean up spilled food products and keep all stored products tightly sealed.

If an insecticide is required, apply a spot treatment according to label directions.

SILVERFISH AND FIREBRATS

Silverfish and firebrats may cause damage in the home by eating foods or other materials that are high in protein, sugar or starch. They eat cereals, moist wheat flour, paper on which there is glue or paste, sizing in paper and bookbindings, starch in clothing and rayon fabrics.

Silverfish and firebrats are common in homes. The silverfish lives and develops in damp, cool places. Large numbers may be found in new buildings in which the newly plastered walls are still damp. The firebrat lives and develops in hot, dark places, such as around furnaces and fireplaces, and in insulations around hot water or steam pipes.

In apartment buildings silverfish and firebrats follow pipelines to rooms in search of food. These insects may be found in bookcases, around closet shelves, behind baseboards, windows or door frames.

Both of these insects are slender, wingless and are covered with scales. Adults are about 1/3 to 1/2 inch long. Silverfish are shiny and silver or pearl-gray in color. Firebrats are mottled gray. The young insects look like adults except they are smaller.

Both insects have 2 long slender antennae attached to their heads and 3 long tail-like appendages at the hind end. Each appendage is almost as long as the body.

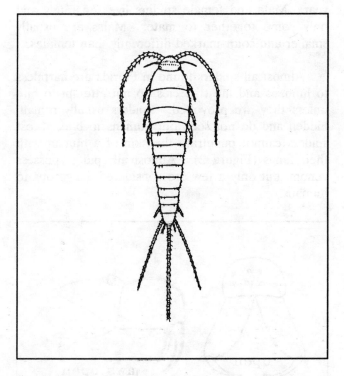

Figure 2. Silverfish.

Silverfish and firebrats are active at night and hide during the day. When objects under which they hide are moved, they dart about seeking a new hiding place.

Biology

Under usual house conditions, silverfish and firebrats develop slowly and have few young. They are hardy and can live for several months without food.

Females lay eggs the year around in secluded places, such as behind books or on closet shelves. Eggs are occasionally laid in the open. Silverfish lay only a few eggs at a time but may lay several batches over a period of weeks. The eggs are whitish, oval, and about 1/32 inch long. They hatch in 2 to 8 weeks, depending on temperature.

Firebrats lay about 50 eggs one at a time and will lay several batches. The eggs are soft, white and opaque when laid. They will later have a yellowish tinge. Firebrat eggs hatch in about 2 weeks. Both insects reach maturity in 3 to 24 months. Their rate of growth depends on temperature and humidity.

Control

Residual sprays are effective for controlling these insects.

The sprays should be applied to floors and wall moldings, behind drawers, under furniture, cracks and crevices, and the floor and ceiling of attics. Control may not be immediate because hiding insects must come out and contact spray residue. Ten days to 2 weeks may be required to determine whether or not control has been achieved.

Dusts of the recommended materials may be used for treating walls, voids, crawl spaces, and attics. Space sprays are useful for controlling exposed insects.

Household Spiders

There are about 1000 species of spiders in the United States. They live everywhere, including homes and buildings. Some species are able to bite humans and inject a venom into the skin. The brown recluse and the widow spiders are considered venomous spiders; however, most spiders are not harmful to man.

BIOLOGY AND DESCRIPTION

All spiders have 8 legs and 2 body regions (Figure 1). They are predators, feeding primarily on insects and other arthropods. When feeding, spiders inject a digestive fluid into their prey, then suck up the digested food. They can survive for long periods of time without feeding. Some spiders have been kept alive for over 2 years without feeding.

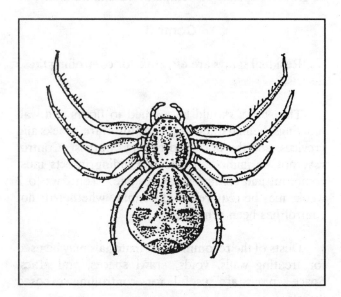

Figure 1. Spiders have eight legs and two body regions (Crab Spider actual size 1 1/2").

Because spiders feed entirely on living insects or other animals, they may actively search for their prey, hide and wait for them to pass, or build webs to trap flying insects. Most web-spinning spiders build and abandon several webs per year. The webs are produced by glands on the spider's abdomen. The silk is a liquid which hardens when exposed to the air.

Silk is used to construct webs, safety lines, egg sacs and as parachutes for long distance travel.

Spiders reproduce by laying eggs in a silken egg sac. The egg sac is either carried around by the female or hidden in the web. Egg sacs of large spiders may contain several hundred eggs. The eggs hatch in about 2-3 weeks after they are deposited. Most young spiders mature to adults in about one year. Male and female spiders live separately and only come together to mate. Males are usually smaller and color-marked differently than females.

Almost all spiders found in Florida are harmless to humans and most species do not attempt to bite unless they are provoked. Spiders usually remain hidden and do not seek out humans to bite. Most spiders cannot penetrate the skin of a human with their fangs (Figure 2). Almost all spiders possess venom, but only a few are considered dangerous to humans.

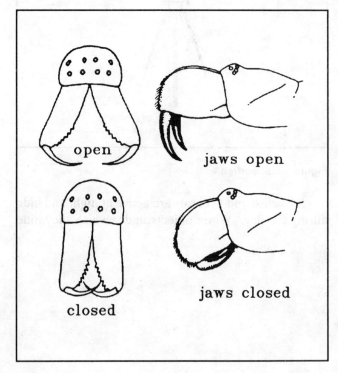

Figure 2. Fangs of the most common spiders.

Spiders are of interest since some invade homes, others are considered poisonous, and some larger species are raised as pets.

HOUSE SPIDERS

Several species of spiders enter houses and become a nuisance. Many people simply dislike spiders and cannot tolerate their presence. When numerous, spiders are annoying because they construct webs. Abandoned webs collect dust, resulting in cobwebs. However, spiders are considered beneficial because they feed on insect pests and other spiders.

Newly hatched spiders are tiny and easily enter homes through screens or around loose fitting doors and windows. Careful screening will keep larger spiders out of homes. If insects they eat are not plentiful, spiders are less likely to infest a home.

Non-chemical control of spiders is usually quite effective in reducing spider populations. Outside lights should not be left on at night. Large numbers of flying insects attracted to lights cause spiders to congregate around garages and under eaves. Trash, lumber piles, bricks, weeds and outside structures are good breeding places for spiders and should be cleaned up. Inside the home, spider webs should be brushed down. The egg sacs should be destroyed to prevent hundreds of young spiders from emerging. Vacuum cleaner attachments may be used to clean walls and the collected debris should be destroyed.

Chemical control of spiders is difficult outdoors because web-spinning spiders do not tend to contact treated surfaces.

Inside the house, space sprays containing pyrethrins or pyrethroids are effective in killing spiders. Space sprays have little residual activity and should be applied when spiders are noticed. When spraying enclosed areas, care should be taken so spiders agitated by the spray do not drop onto the person doing the spraying. Repeat the treatment in 4 weeks to kill recently hatched spiders.

Dust formulations can be used in crawl spaces, attics and utility areas to provide long-term protection. Insecticidal dusts tend to cling to the spider webs for long periods of time. When spiders chew their webs to recycle the silk they consume the toxicant and die.

Springtails

Springtails are extremely small insects that become problems in homes or other structures when they invade in enormous numbers. They are very common outdoors and are our most common soil insects. As a result, they fulfill an important role in soil development and enrichment. There may be as many as 50,000 per cubic foot in forest litter. They can also occur indoors in potted plants and decaying bulbs.

DESCRIPTION AND HABITS

These insects are white or grey in color. They have a forked appendage to the rear and bottom of the abdomen. Used as a lever, this appendage, allows these insects to jump or spring into the air. And this is how these insects received their name.

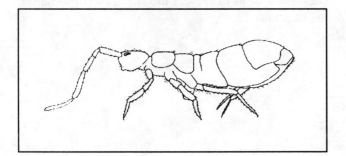

Figure 1. Springtail

They infest buildings that have constant high humidity. This is usually in the basement, but may be in other areas that have water leaks. As a result, the best method of control is to stop the leak or decrease the humidity in the building. Fans may be used to quickly dry out wet areas.

Springtails are attracted to light and may enter homes or other structures under the doors. They can invade buildings from infestations in mulch or other moist breeding areas.

Crickets

Crickets are sometimes nuisances in buildings and they may also damage fabrics or other materials. They are especially destructive to silks and woolens. They are also attracted to perspiration and other stains on clothing and fabrics. Occasionally crickets invade a structure in large numbers. They are often attracted to lights around a building at night. Besides the damage they may cause, they produce a chirping sound which may, after a period of time, annoy inhabitants.

Crickets belong to the insect order Orthoptera and are related to grasshoppers. These insects do not undergo a complete metamorphosis, therefore, the young resemble adults except they do not have functional wings. Young and adults both have similar feeding habits.

The most common crickets to invade buildings include the house cricket, *Acheta domesticus*, and the field cricket, *Gryllus* spp., which are very similar in appearance.

House cricket adults range in length between 1/2 and 3/4 inch. They may be light yellowish brown, with three dark bands on the head, or solid shiny black. This species has long, slender antennae. The field cricket is slightly larger, up to 1 inch in length, and usually brown or black. Females of both species have a long, thin ovipositor projecting from the tip of the abdomen.

MANAGEMENT GUIDELINES FOR CRICKETS

The key to managing crickets in buildings is exclusion. Cracks and other openings from the outside that provide access to the building should be sealed. Caulk or otherwise seal cracks and crevices inside the building that provide hiding places. Behind or under heavy furniture and appliances or in other inaccessible areas, it may be possible to remove crickets using a strong vacuum cleaner. Weeds and debris around the outside of the building should be removed to eliminate attractive habitats. Change outside lighting to sodium vapor lights or yellow incandescent lights that are less attractive to crickets

(and other insects). Garbage and other refuse that serves as food should be stored in containers with tight lids and elevated off the ground on platforms or bricks.

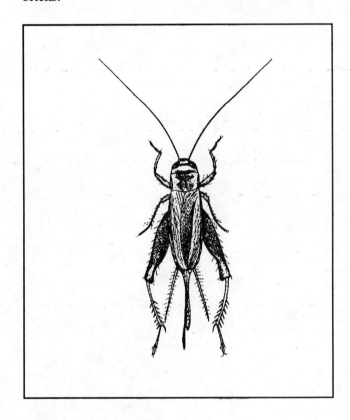

Figure 1. House cricket

Insecticides should be used only when exclusion and sanitation cannot accomplish control quickly enough to stop the damage within a reasonable time. Use liquid sprays of an insecticide registered for use indoors as a spot spray in cracks and crevices and other areas where crickets may hide. Sorptive powders may also be blown into inaccessible areas. Apply liquid sprays around the perimeter of the building or in other outdoor areas if crickets cannot be controlled through sanitation. Avoid using outdoor spray materials indoors unless the label states that this is permissible. Insecticide impregnated baits or granular formulations of certain materials may also be used outdoors around buildings for cricket control. Granules are suitable in lawns and other areas subject

152

to moisture or frequent watering. Avoid the use of baits or granules if children or pets can gain access to them.

Cricket infestations are usually seasonal. Most often problems occur during the fall as evenings become cooler and the insects seek buildings for warmth and shelter. Because of this, applications of long-residual insecticides are not usually needed indoors for adequate control.

Rats and Mice

Rats and mice are important rodent pests entering Florida homes and warehouses for food and harborage. These rodents eat any kind of food that people eat. They also contaminate 10 times as much food as they eat, with urine, droppings and hair. They can carry at least 10 different kinds of diseases including bubonic plague, murine typhus, spirochetal jaundice, Leptosporisis, rabies, ratbite fever and bacterial food poisoning. Many times rats bite sleeping children while trying to get bits of food off them that were not washed off before bedtime. Rats and mice also start fires by gnawing matches and electrical wires in homes.

The Norway rat, roof rat and house mouse are the most persistent rodent populations in need of control.

Figure 1. Roof rat (top), Norway rat (middle) and house mouse (bottom).

NORWAY RATS

In Florida, Norway rats are most common along the sea coasts and canals. They thrive particularly in areas where garbage is not properly stored. Although Norway rats generally prefer to eat fresh meat, fish, and grain, they can survive quite well on an ounce per day of garbage or decayed food along with an ounce of water. Frequently they range 100-150 feet from harborages in search of food or water.

Norway rats are burrowers and often dig in rubbish and under buildings or concrete slabs. Burrowing can cause damage by undermining the foundations of buildings, eroding banks of levees, disfiguring landscape plantings, and blocking sewer lines.

This rat is reddish-brown and heavy-set with a blunt muzzle. Its tail is about as long as the combined head and body. Adults weigh 3/4 to 1 pound. Their droppings are 3/4 inches long and capsule-shaped. Norway rats live about 1 year and reach sexual maturity in 3-5 months. They have 8-12 young per litter and up to 7 litters per year.

ROOF RATS

Roof rats thrive in attics, roof spaces, palm trees and ornamental shrubbery. They are climbers and prefer to nest off the ground. Roof rats are destructive to citrus groves, since they live in citrus trees and gnaw on the fruit. They can be quite destructive in attics, gnawing on electrical wires and rafters.

Roof rats generally prefer vegetables, fruits and grain, and consume 1/2 to 1 ounce per day of food from various sources. Because they must have water to survive, roof rats also consume an ounce per day and will range 100-150 feet from harborages in search of this and food.

Color ranges from black to grizzled gray to tan with a light belly. The tail is longer than the combined head and body. Adults weight from 1/2 to 3/4 pound. Their droppings are up to 1/2 inch long and spindle-shaped. Roof rats live about 1 year and reach sexual maturity in 3-5 months. They have 6-8 young per litter and up to 6 litters per year.

HOUSE MICE

House mice normally live outdoors in fields, occasionally migrating into structures. In houses, they live behind walls and in cabinets and furniture.

They prefer to feed on grains but usually nibble at a wide variety of foods. House mice require only 1/10 ounce of food and 1/20 ounce of water daily, but can survive on food alone if it has high moisture. Frequently house mice range 10-30 feet from harborages.

House mice are brown to gray in color with the tail as long as the body. Adults weigh about 1/2 ounce. Their droppings are 1/8 inch long and rod-shaped. House mice live about 1 year and reach sexual maturity in 6 weeks. They have 5-6 young per litter and up to 8 litters per year.

Important Rat Behavior

The movement of rats and mice is usually related to food, water or harborage. Knowing where they are likely to go is important to controlling them.

Rats use any method to get to food, water or harborage. Their excellent sense of balance enables them to run on pipes, narrow ledges and utility wires. Rats, especially roof rats, will climb anything their claws will hold on to, including wires, pipes and rough walls. Because rats are excellent swimmers, they often live in sewers and occasionally enter homes through toilets.

Rats like to use regular paths or runways along walls or behind debris. To get food in the open, they will run behind things to get as close to the food as possible. They are afraid of strange objects or strange food and may avoid both.

Norway and roof rats are both aggressive species. The Norways are usually more aggressive, driving roofs from the territory. Both species are seldom found in the same building.

Rats and mice frequently gnaw on their surroundings. Their teeth grow 4-1/2 to 5-1/2 inches per year and only gnawing keeps them short and sharp.

Rats and mice are active mostly at night. Rats show greatest activity the first half of the night, if food is abundant. Mice usually are active at night both right after dark and between midnight and dawn. Both rats and mice will be active during daytime hours when food is scarce, when there is an overpopulation of rats, or when a poison has been used and the population is sick.

Recognizing Rat and Mouse Signs

Since rats and mice are active at night and are rarely seen during the day, it is necessary to recognize signs of their activity.

Droppings and Urine

Most people first recognize rodent problems by finding droppings or urine stains in and around buildings. Rodents usually have favorite toilet areas but will void almost anywhere. Old droppings are gray, dusty and will crumble. Fresh droppings are black, shiny and puttylike. Rodents urinate while running, and streaks are characteristic. The urine glows under ultraviolet lights and glows blue-white when fresh.

Figure 2. Droppings of roof rat (1/2", left), Norway rat (3/4", middle) and house mouse (1/8", right).

Gnawed Objects

Rodents gnaw every day in order to keep their teeth short and sharp. Rats also gnaw to gain entrance or to obtain food. Teeth marks on food, building materials, wire, and edges of beams are indications of gnawing. Rodents will gnaw holes in wooden walls, pressed wood and posts. Fresh gnawing in wood is usually light-colored with sharp, splintery edges. Old gnawing is smooth and darker.

Runways

Rats habitually use the same paths or runways between harborage and food or water. Outside runways are paths 2-3 inches wide and appear as smooth, hard-packed trails under vegetation. Indoors, runways are usually along walls. Undisturbed cobwebs or dust indicates runways are not being used.

Rubmarks

Along runways, dark greasy rubmarks appear from contact with the rodent's body. Rubmarks on walls appear as black smudges left by the rodent. New rubmarks are soft and will smudge. Old rubmarks are brittle and will flake when scratched. Rafters may show swing marks of roof rats.

Tracks

To detect rodent activity, spread dust material like talcum powder along runways. Footmarks of rats (5-toe hind foot, 4-toe front foot) or tail drag marks will show in the powder.

Burrows

Norway rats burrow for nesting and harborage. Burrows are usually found in earth banks, along walls, under rubbish and concrete slabs. Freshly dug dirt scattered in front of 3-inch openings with runways leading to the openings is characteristic. Burrows usually are 18 inches deep in most soils. Slick, hard-packed runways indicate an old, established colony.

Live Rats and Dead Rats

The sighting of live rats is a sure sign of infestation. Sightings in the daytime indicate large populations, disease or poisoning. Mummified rat carcasses may indicate a former infestation although many fresh carcasses suggest disease or poisoning.

Sound

Usually rodent sounds are heard at night or in quiet areas. Rodents moving at night often scratch, gnaw, and fight. The young often squeak while in the nest.

Rodent Control

To control rats, implement the following procedures:

Rodentproofing

Rodentproofing is changing the structure of buildings in order to prevent entry of rats and mice. In considering rodentproofing, you must know that
- Rats can squeeze through cracks 1/2 inch wide; mice, 1/4 inch wide. Any place a pencil can be poked, a mouse can go.
- Rats can climb the inside of vertical pipes 1-1/2 - 4 inches in diameter.
- Rats can climb the outside of vertical pipes up to 3 inches in diameter and any size if within 3 inches of a wall.
- Rats can jump vertically 36 inches, horizontally 48 inches, and reach horizontally or vertically 15 inches.
- Rats can jump 8 feet from a tree to a house if the branch is 15 feet above the roof.

Rodentproofing requires the use of rodent-resistant materials. The following materials are considered rodent resistant:
- Sheet metal (26 gauge or heavier).
- Perforated metal (24 gauge or heavier with openings no more than 1/4 inch).
- Hardware cloth (19 gauge or heavier with openings no more than 1/4 inch).
- Brick with mortared joints.
- Cement mortar (1:3 mixture).
- Concrete (1:2:4 mixture).

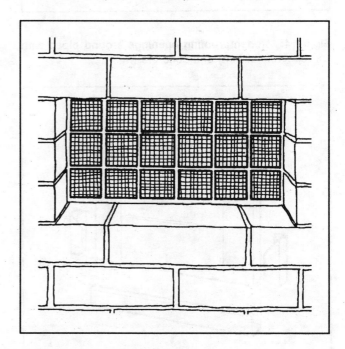

Figure 3. Rodentproofing a vent.

A gnawing edge is the edge of substances which rats can gnaw through. The gnawing edges must be

156

protected with rodent-resistant materials. Places to rodentproof are edges of doors, windows, holes where pipes enter buildings, ventilation holes in foundations, roof vents, exhaust fans, and eave vents. Rodents can also enter homes through toilets.

To be effective, rodentproofing must block all possible rodent entry points. During the first 2 weeks after searching rodents will find breaks in the rodentproofing. Inspect frequently during this time and promptly repair any breaks. Eliminate the rodents trapped indoors due to rodentproofing.

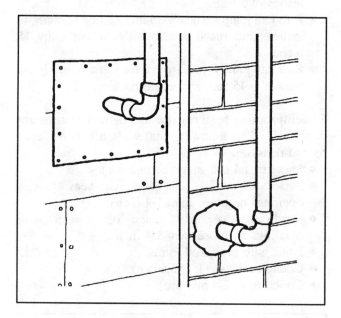

Figure 4. Rodentproofing openings around pipes with sheetmetal (left) and concrete (right).

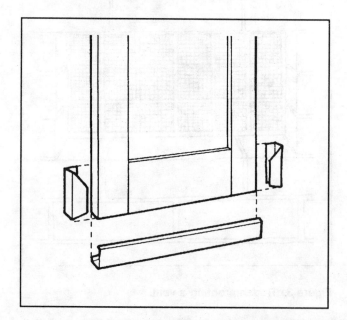

Figure 5. Rodentproofing a door, placing channel at bottom and cuffs at sides over channel.

Sanitation

Good housekeeping or sanitation is a basic factor in rodent control. Eliminating food, water and harborage for rats and mice can reduce rodent populations rapidly. To implement sanitation practices:
- Clean up garbage and rubbish.
- Properly store garbage (metal garbage cans should have tight-fitting lids).
- Properly store food (store raw or prepared foods and refuse indoors in covered, ratproof containers or in ratproof rooms).
- Store pet food and bird seed in ratproof containers.
- Remove harborages (remove piles of rubbish, trash, junk, boxes and protected enclosures).
- Dry up sources of water.
- Pick fruits and vegetables when ripe so rodents will not feed on them.

Sanitation must be used constantly if rodent control is to be effective. Yearly cleanup programs are generally ineffective for rodent control.

Predators

Many people have relied on cats and dogs to control rats, but in general cats and dogs are not good tools for control. Food put out for pets is excellent rat food. Most people put out more food than the pet can consume in one day. Rats then clean the bowl overnight. Because pets are well-fed, they are too lazy to hunt. Studies have shown that although predators can keep an area rat free, they cannot remove an existing infestation.

Birds of prey, hawks and owls, feed on large numbers of rodents. Barn owls are exceptional rat killers and a pair can be expected to kill several hundred rats over a one year period.

Many species of non-poisonous snakes are very beneficial in rodent control. Snakes such as rat snakes, king snakes, pine snakes, black racers and coach whips eat numerous rodents and are important in controlling rodent populations. Do not kill non-poisonous snakes.

Trapping

Trapping is an underrated method of controlling rodents. One reason trapping is often overlooked is that snap traps have been around for a long time and are cheap. Traps can be used to eliminate rats where poison baits would be dangerous, to avoid dead rat odors and to eliminate bait-shy rats.

It is important to place traps where the rats are. Rats and mice are used to human odors so there is no need to use gloves when handling traps. Since mice travel only 10-30 feet but rats travel 100-150 feet from harborages, more traps are needed to trap mice than rats in a house.

Rats and mice also have different behavior around new objects. Rats are cautious, and it may be a week before they approach a trap. Mice are curious and will normally approach traps the first night. If you don't catch a mouse in the first few nights, the trap is in the wrong location. To help rats overcome trap shyness, place traps unset, in place, for several days. This allows rats to overcome shyness and results in better catches.

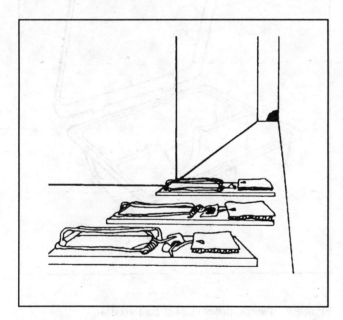

Figure 6. Traps at right angles to rat run.

Baited traps rely on the rat's being attracted for feeding. The bait must compete with other available foods, so no one bait is ever the best bait for all locations. Rodents living on garbage or spoiled food prefer something fresh. The following are some baits that have proven to be successful:

- Whole nuts for rats and mice.
- Raisins or grapes for roof rats.
- Sardines packed in oil for Norway rats.
- Peanuts or peanut butter for rats and mice (soak whole peanuts in water overnight; old peanut butter becomes rancid so replace it frequently).
- Dry rolled oatmeal is excellent for mice.
- Bacon squares.

- Small wads of cotton for mice and rats (they look for nest material).
- Gumdrops for mice.

Baited traps should be set a right angles to rat runs. Traps can be nailed to rafters and beams to take advantage of areas where rats travel. Set traps along walls, behind furniture, and near holes. Remember to set traps where children and pets will not be hurt.

Runway traps

These traps catch rats when they accidentally bump the trigger. Runway traps are available or can be made from snap traps by enlarging the trigger with cardboard, hardware cloth or screening. There is no bait to go stale, so there is an increased chance of success. In placing runway traps, the trap should be placed at right angles to the wall or along runways. To hold the trap in place on pipes or rafters, use rubber bands, nails, or hose clamps.

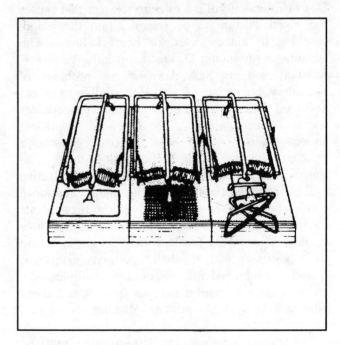

Figure 7. Runway traps made from enlarged snaptraps.

Glue boards

Special glue can be placed in pie tins or paper plates. The glues do not harden but will hold a rat in place. Other rats become curious and also get caught. Placing a small piece of bait in the center of a glue board can increase effectiveness. Dusty and wet conditions will impair the trap's effectiveness. Glue boards are better suited for mice and safe for children and pets. Boards may be cleaned with cooking oil.

Poison Baits

Traps are effective usually when dealing with small numbers of rats. When rats are plentiful or where unsanitary conditions exist with harborage, poisoned baits are the best control method.

Poison baits are available as ready to use, premixed baits. They come in many forms: parafinized blocks for outdoor use and high humidity areas; treated meal, seeds, or parafinized pellets in bulk; and in "place packs" for indoor use. Water baits are sold as packets of concentrate that are mixed with water. They are administered with a chick waterer and are useful in areas where rodent food is abundant.

There are three types of rodenticides: acute toxins, calcium releasers and anticoagulants. Most acute toxins are no longer avilable due to the risk of accidental poisoning. One new acute toxin that is considered safe and effective is bromethalin. Vitamin D_3 or cholecalciferol is a calcium releaser that causes too much calcium to be released into the blood, resulting in kidney, liver, or heart failure. The advantages of vitamin D_3 are that it kills Warfarin®-resistant rodents and there is no problem of secondary poisoning of pests or wildlife that eat poisoned rodents. The most common rodenticides are the anticoagulants. The older are multidose anticoagulants. The new second generation anticoagulants are effective after a singe dose. The single-dose anticoagulants are generally effective against rodents resistant to the older multidose compounds. Rodents poisoned with anticoagulants die from internal bleeding. Because the effects show up days after taking the bait, they do not associate their poisoning with the bait. Anti-coagulants are considered safer rodenticides, because multiple doses are necessary to poison humans or pets and these poisons have a simple antidote, Vitamin K.

Whenever a rodenticide is used, safety must be the first consideration. Poison baits should be placed where they are inaccessible to children and pets. Where rodent runs are exposed and in most outdoor situations, tamper proof bait boxes should be used. Baits should always be placed where rats live and breed.

About 1 pound of anticoagulant bait should control most rats in and around homes. The baits should be placed in stations with 1/4 pound of bait per station. Shallow containers for holding the bait are best. For added effect, water may be provided separately for the rats to drink. Paraffin blocks of baits may be placed outdoors where rodents feed. Pick up dead rats wherever they are noticed. A few cases of pet poisoning have been reported when pets feed on dead rats or mice. When rats die in areas where they can not be removed, it may be necessary to ventilate the area or use odor absorbent or masking products. Usually anticoagulant poisoned rodents are dehydrated and do not produce severe odor after death.

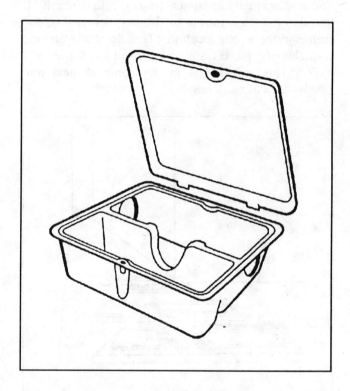

Figure 8. Tamperproof rodent bait station.

When you control your rats, encourage your neighbor to control his/her rats at the same time. The greater the area that is controlled, the more effective the results will be and the longer it will take new rats to migrate back.

Check with your local county health department to determine whether a rodent control project is active in your neighborhood. The department may be able to offer advice and aid in controlling rats. Remember it is no disgrace to acquire some rats, but it is a disgrace to maintain them.

Non-Chemical Rodent Control

INTRODUCTION

Rats and mice often enter homes, farm buildings and warehouses in search of food and shelter. The most common rodent pests in Florida are the commensal rats and mice. These are Old World rodents that have adapted to live with man. They include the roof rat, Norway rat, and house mouse. These commensal rodents have been carried by man to every corner of the Earth. Rats and mice consume or contaminate large quantities of food and damage structures, stored clothing, and documents. They also serve as reservoirs or vectors of numerous diseases, such as ratbite fever, leptospirosis (Weil's Disease), murine typhus, rickettsial pox, plague, trichinosis, typhoid, dysentery, salmonellosis, *Hymenolepis* tapeworms, lymphocytic choriomeningitis, and rabies.

In most cases of rodent infestation, the pest animals can be controlled without having to resort to the use of poisons. Good sanitation and exclusion will prevent most problems. If rodents do find their way indoors, small populations can be easily eliminated with various nontoxic methods. Rodenticides (rodent poisons) need only be used in cases of large or inaccessible infestations. The trapping of rodent pests is often preferable to the use of poisons. Traps prevent rodents from dying in inaccessible places and causing an odor problem. There is no chance of an accidental poisoning or secondary poisoning of nontarget wildlife, pets or children with the use of traps. Secondary poisoning of pets or wildlife can result from eating poisoned rodents. Traps can be used in situations where poisons are not allowed, such as in food-handling establishments.

RODENT ECOLOGY
"KNOW YOUR OPPOSITION"

The house mouse is the most common commensal rodent invading houses in Florida. It is primarily nocturnal and secretive. The presence of mice is usually indicated by sightings, damage from gnawing into food containers or presence of droppings. In the wild, house mice feed primarily on seeds. In the home, they prefer grain products, bird seed and dry pet food. Peanut butter or gum drops stuck to the trigger and rolled oats or bird seed sprinkled on the trap are good baits. House mice are inquisitive and actively explore anything new. They tend to nibble on many small meals a night. House mice are good climbers. They have a small home range and usually stay within 10 to 30 feet of their nest. Therefore, traps for mice should be set 6 to 10 feet apart. Nests are usually in structural voids, in undisturbed stored products or debris, or in burrows outdoors. When food is abundant, nesting material, such as a cotton ball, tied to the trigger can act as an effective lure. Mice and rats are very nervous about moving in the open. The more cover they have, the more comfortable they are. They would prefer running behind an object or along the baseboard of a wall than running across an open space.

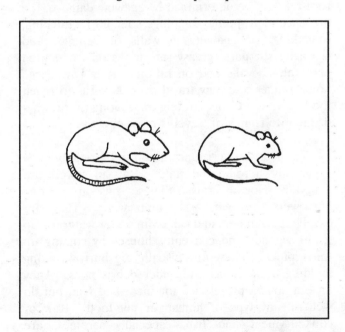

Figure 1. Left: young rat (6-7"); feet large, head large
Right: House mouse (6-7"); feet small, head small.

The roof rat or black rat is the most common rat in Florida. These rats are excellent climbers and often nest in attics, wall voids, hollow trees and palm thatch. They prefer to travel off the ground and enter

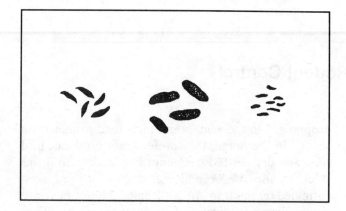

Figure 2. Droppings, L-R: Roof rat (pointed, average length 1/2"); Norway rat (blunt, average length 3/4"); House mouse (pointed, average length 1/8").

the ground and enter houses from nearby trees or along powerlines. Roof rats prefer fruit (they are sometimes called citrus rats), but will eat any type of human, pet or livestock food. Peanut butter, pieces of fruit or nut meats are the best baits. Rats are usually fearful of new items in their environment and avoid them for several days. This means that traps should be left in place for at least one week before they are moved to a new location. The presence of roof rats can be determined by gnawing damage, the presence of droppings, sightings, sounds of scratching, squeaking, or gnawing in walls or ceilings, and characteristic dark, greasy rub marks on frequented paths along walls and on rafters. Rats have large home ranges and may travel over 50 yards to reach food or water. Concentrating traps along rat runways or favorite routes of travel is most effective.

The Norway rat is uncommon in Florida, but can occur anywhere in the state. Rats in sewers are generally Norway rats. These rats are strong burrowers, but can also climb well. They are excellent swimmers and can swim under water for up to 30 seconds and can enter houses by coming up toilet pipes. These rats usually dig burrows along building foundations and under debris piles. They have a strong preference for meat and fish, but do well on any type of human or pet food. Raw or cooked meat and fish, especially sardines, are excellent baits, but peanut butter also works well. Like the roof rat, the Norway rat is cautious of new objects and has a very large home range, over 50 yards in radius. The Norway rat is very aggressive and will drive roof rats out of an area. However, both species of rats can be found in the same

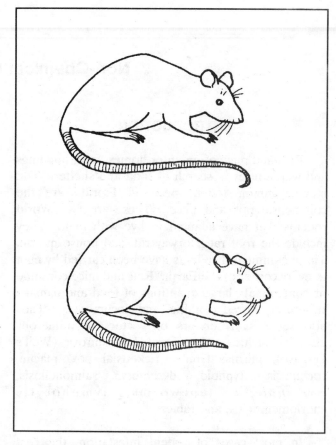

Figure 3. Top: Roof rat (12-17"); tail longer than head and body; body light and slender; ears larger. Bottom: Norway rat (12-18"); tail shorter than head and body; body heavy and thick; ears small.

building, with roof rats in the attic and Norway rats in the basement.

SANITATION AND EXCLUSION

Proper sanitation will do a great deal to control rodent pests. All animals have three requirements for life: food, water, and cover. Removal of any one will force an animal to leave. The removal of debris such as, piles of waste lumber or trash, used feed sacks, abandoned large appliances, and trimming the dead fronds from palm trees will substantially reduce the harborages for rodent pests. Stacked firewood stored for long periods provides good harborage for all three commensal rodents. Storage of pet food and seeds, such as wild bird seed, in rodent proof containers of glass or metal, eliminates these food sources. Collect and remove fallen fruit from backyard trees and orchards. Keep lids on trash cans and close dumpsters at night, making areas less attractive to rats and mice. The drainage holes in dumpsters should be covered with hardware cloth to keep rodents out.

Exclusion is also called rodent proofing. This involves making your home a fortress that rodents cannot breach. Rodents can squeeze through any opening that their heads can fit through. That is a 1/4 inch opening for mice and a 1/2 inch opening for young rats. Young rats and mice are the dispersing individuals, so these are the ones most likely to invade new areas, like your home. Any opening that a pencil can fit through will admit a mouse. Below is a list of recommended materials for excluding rats and mice.

1. Galvanized, stainless, or other non-rusting metal.
   * Sheet metal, 24 gauge or heavier.
   * Expanded metal, 28 gauge or heavier.
   * Perforated metal, 24 gauge or heavier.
   * Hardware cloth, 19 gauge or heavier, 1/4 inch or smaller mesh.
2. Cement mortar with a 1 part cement: 3 part sand mix or richer.
3. Concrete with a 1 part cement: 2 part gravel: 4 part sand mix or richer. Broken glass added to mortar or concrete will deter rodents from tunneling through a patched hole before the material hardens.
4. If in good repair, brick, concrete block, tile or glass will exclude rodents.
5. Wood will exclude rodents if no gnawing edges are present.

TRAPS

There are four main types of rodent traps: snap traps, multicatch traps, single catch live traps, and glue board traps. Snap traps include the classic rodent traps with the wood base and the newer metal clothespin traps. They are designed to kill the trapped animal quickly and humanely. Snap traps should not be set where children or pets will come in contact with them. There are three different types of triggers: wood / prebaited, metal for holding bait, and expanded trigger, which is used in runways. The expanded trigger is the most versatile since it also can be baited. Older snap traps with other types of triggers can be modified to produce an expanded trigger.

Traps should be placed where rodents are likely to be. Rodents are creatures of habit and prefer to follow the same runways. It is important to identify these and place traps there. Runways can be identified by sprinkling a fine layer of flour or baby powder in suspected areas and looking for tracks. This is a safe diagnostic method for determining rodent activity, but should not by confused with the use of rodenticide tracking powders that require a restricted-use pesticide license. Rodents often run along edges and traps should be set along walls, especially where objects such as a box or appliance will guild them into the trap. The type of bait used depends on the species of rodent pest. Roof rats prefer to travel above the ground and are easier to trap along these precarious pathways than on the ground.

Multicatch traps are designed to repeatedly catch a rodent and reset themselves for another capture. Advantages of these traps are the ability to capture several rats or mice with one setting and the scent from the captured mice entices others to the trap. The disadvantages are that the captured mice or rats are alive and must be dealt with, and these traps are expensive. Methods for dealing with the captive rodents includes submerging the entire trap in a bucket of water and drowning them, using drowning attachments available for some traps, placing glue boards in the holding compartment of the trap, or giving the rodents to a snake owner. The release of commensal rodents outside is not a solution, since they will quickly find a way back into your home or someone else's. Trap-wise rodents are also more difficult to trap than naive ones. Multicatch traps must be checked on a regular basis like any other trap to prevent the capture rodents from starving or dying of thirst and creating an odor problem. Available multicatch traps include the Kness "Ketch-All" Automatic Mouse Trap, the Victor Tin Cat Repeating Mouse Trap, and the "Rat Katcher" (previously the "Katch-All") Repeating Rat Trap.

Single-catch live traps are rodent-sized cage traps of various styles. These traps capture the rat or mouse alive and unharmed, but again you have to deal with the captured rodent. The native rodents, cotton mice and eastern wood rat, that occasionally invade rural and suburban homes can be released

162

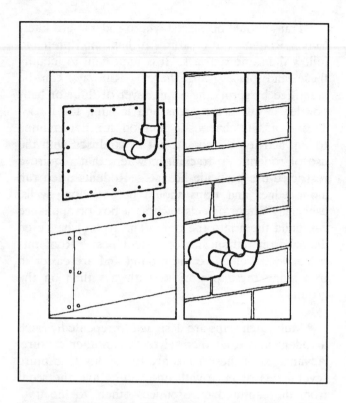

Figure 4. Rodentproofing openings around pipes with sheet metal (left) and concrete (right).

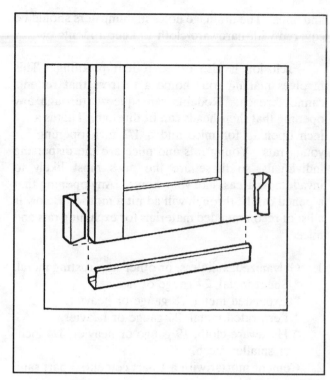

Figure 6. Rodentproofing a door, placing sheet metal channel at bottom and cuffs at sides, over channel.

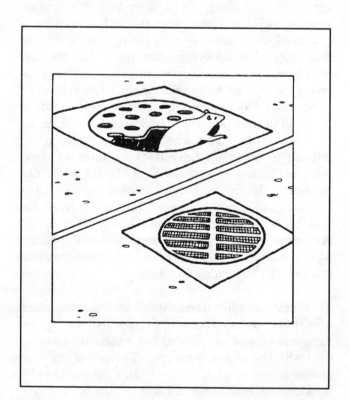

Figure 5. Rodentproofing drains with 1/4" hardware cloth.

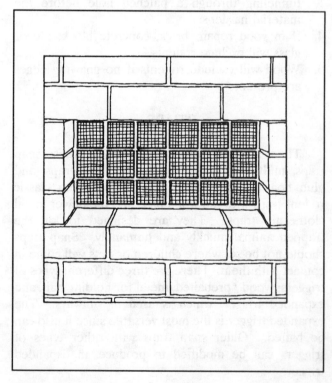

Figure 7. Rodentproofing a vent with 1/4" hardware cloth.

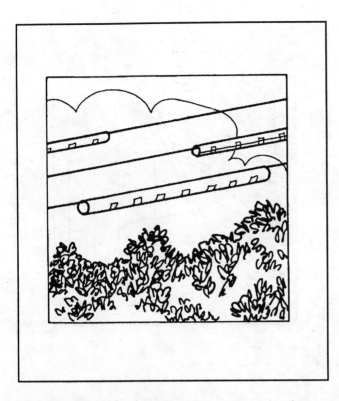

Figure 8. Rodentproofing utility wires to limit access to buildings using rolling plastic tubes made from rectangular sheets of plastic. The tube rolls when the rodent tries to walk over it.

Figure 10. Rodentproofing air vents and chimneys using 1/4" hardware cloth.

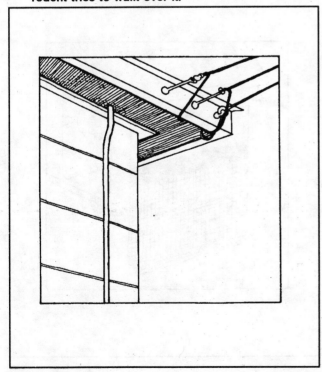

Figure 9. Rodentproofing openings where wires enter buildings.

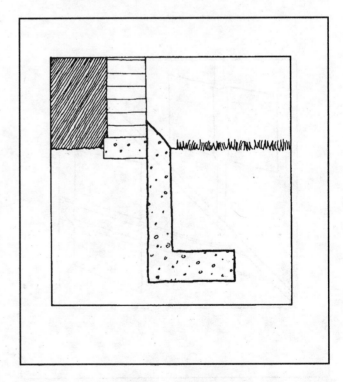

Figure 11. Foundation curtain wall should extend at least 2 ft. below ground level and horizontal lip should be at least 1 ft. wide. Thickness should be at least 4 inches.

164

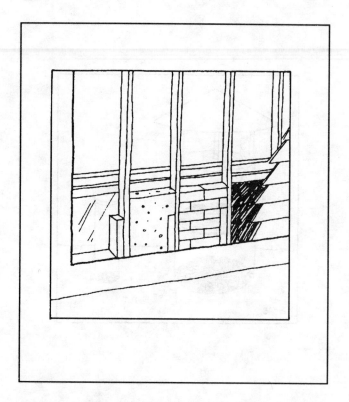

Figure 12. Blocking end spaces of wall void using sheet metal, concrete, brick, or wood.

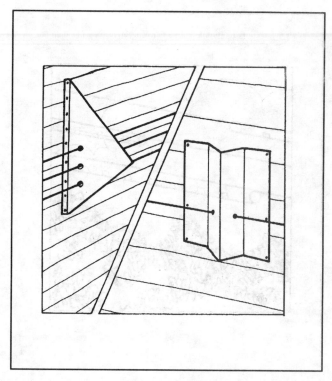

Figure 14. Rat guards for utility wires near a wall.

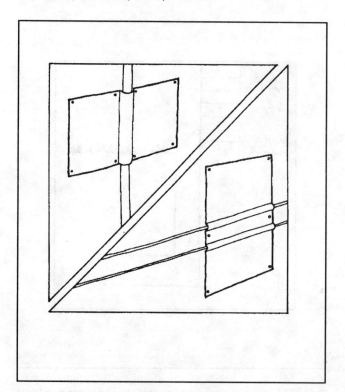

Figure 13. Rat guard over pipes and utility wires against a wall.

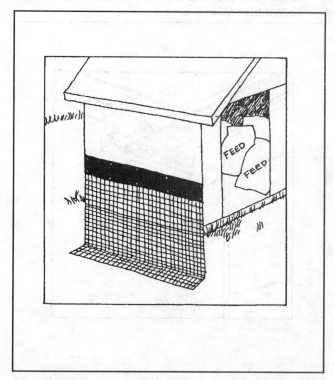

Figure 15. Hardware cloth curtain wall on an storage building. Top edge covered with strip of sheet metal.

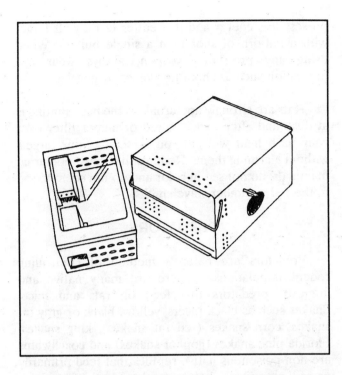

Figure 16. Multicatch Mouse Traps - (Left) Victor Tin Cat;   (Right) Kness Ketch-All.

back in the woods with little chance of their returning indoors. They can be recognized by their fine brown fur, white belly, large eyes, and very large ears. Commensal rodents should not be released because they will return to your home or someone else's. Rodents caught in these traps are best dispatched by submerging the entire trap in a bucket of water. These traps should be used in areas of Florida known to be occupied by endangered native rodent species, especially on barrier islands and the Florida Keys, to confirm the species of invading rodent and prevent the accidental killing of an endangered species. These traps should be placed against walls or in runways. The most effective bait for mice with this type of trap is rolled oats (uncooked oatmeal) sprinkled inside the trap with a fine trail leading out. Rat-sized live traps are produced by Havahart, Kness Mfg., Mustang Live-catch Traps, Safeguard Live Animal Traps, Sherman Live Traps, and Tomahawk Live Traps. Mouse-sized live traps are produced by Havahart, Sherman Live Traps, Tomahawk Live Traps, and Trap-Ease Mouse Live Trap.

Glue boards are used just like snap traps. While both rat and mouse-sized glue boards are made, these traps are most effective against mice. Rats are often

strong enough to pull themselves free from glue traps. Glue boards should not be set in wet or dusty areas

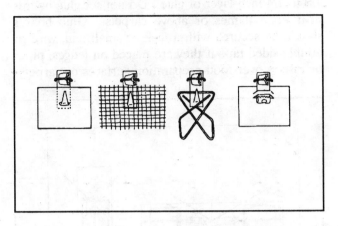

Figure 17. Methods of converting metal bait-triggers to expended triggers for runway sets.

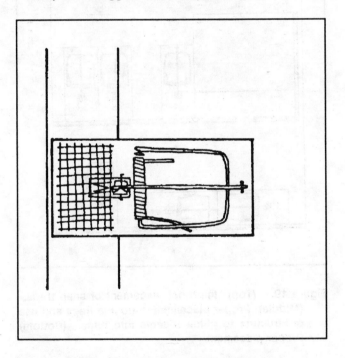

Figure 18. Snap trap placement on pipes or rafters. Secure trap with duct tape, wire, or small nails.

because these conditions render the traps ineffective. Wet feet and fur will not stick to the glue and dust coats the glue until it is no longer sticky. These traps also should not be set where children or pets will contact them. Glue boards are not hazardous to children or pets, but the encounter will create a frustrating mess. Clean hands with room temperature cooking oil and clean surfaces with paint thinner or mineral spirits. The best glue boards have at least a 1/8 to 1/4 inch layer of glue. Do not set glue boards near open flames or above carpets. Glue boards should be secured with a tack or small nail, wire or double-sided tape if they are placed on ledges, pipes or rafters over food-preparation surfaces or carpets.

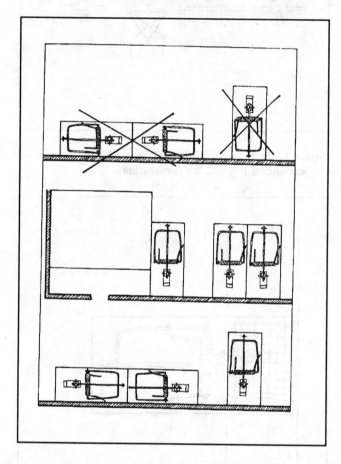

Figure 19. (Top) Improper placement of snap traps. (Middle) Proper placement of double traps and use of structure to guide rodents into traps. (Bottom) Proper placement.

SHOOTING

Shooting rodent pests is not an efficient control method. But, if you choose to use this method, observe the following safety rules. Remember that discharging a firearm within city limits is illegal, as is the use of a firearm by a minor without adult supervision. A .22 cal. bullet can travel over a mile and can easily penetrate corrugated metal walls and roofs, so always be sure of your backstop when using this weapon or any firearm. Shot cartridges are safer than solid bullets because each of the smaller pellets

possess less energy and it is easier to hit your target with a pattern of shot than a single bullet. When using any projectile weapon, always wear eye protection such as shooting glasses or goggles.

Rats are strongly nocturnal, so the best hunting is at dusk and after dark. A red or amber filter over your flash light will aid you in seeing your targets without alarming them. Rodents, like most nocturnal mammals, do not see in color and do not seem to see in the red or amber wavelengths.

PREDATORS

Predators are nature's method of controlling rodent populations. There are many native and domestic predators that feed on rats and mice. Snakes such as black racers, yellow, black or gray rat snakes, corn snakes (red rat snakes), king snakes, Florida pine snakes (gopher snakes) and coachwhips are non-poisonous native reptiles that feed primarily on rodents and may help control outdoor infestations. Hawks and owls, especially barn owls, eat large numbers of rats and mice. Nest boxes of the proper proportion will encourage barn owls and screech owls to nest in your area and raise their young. Hawk and owl parents kill many more rodents when they are feeding their hungry broods. Foxes, bobcats, striped and spotted skunks, weasels and mink will all eat plenty of rodent pests, but these wild predators avoid people. Domestic cats, dogs and ferrets help in controlling rodents in some situations. In general, dogs and cats are most effective at preventing an infestation than eliminating a current population. This is because they are better able to catch and kill an invading rodent that does not know any escape routes, than an established animal that knows numerous escape points. Cats are very effective predators of mice, but usually will not attack an adult rat. They will also kill birds at bird feeders, wild rodents and baby rabbits, so these factors must also be considered. To prevent cats from becoming a pest themselves, be sure to have any cat you release spayed or neutered. This service is required and provided by most county humane societies at the time of adoption. Pet ferrets will kill rats and mice indoors but should never be released outside. The establishment of wild ferret populations could decimate our native wildlife. Many people propose the mongoose for rodent control, but the importing, possession or release of any mongoose is strictly illegal because of the ecological damage they can do. The mongoose has repeatedly shown a preference for native birds and mammals over commensal rodent pests.

ULTRASOUND DEVICES

The principal of ultrasonic devices is to create a loud noise above the range of human hearing (above 18-20 kHz) that is unpleasant to pest species. The problems with ultrasound are numerous. Animals can adapt to most situations, and in a short amount of time they become accustomed to the sound. If the original attractant, such as food, is present, the rodents will return. The short wavelengths of ultrasound are easily reflected and creates sound shadows and the rodents simply shift their activity to these low noise shadows.

Ultrasonic devices will not drive rodents from your home if food, water and shelter are available. However, ultrasonic devices may have a part to play in rodent integrated pest management. Ultrasonic devices may increase trapping effectiveness by altering the normal movement patterns of individual rodents. Traps set in the sound shadow areas will become more effective because the rodents will be concentrated in these areas. The high cost of the units must be considered against the increase in trapping effectiveness to determine if they are cost effective.

Bats

Bats are highly beneficial wild mammals. They are not flying rodents, but belong to a unique order of mammals called the Chiroptera (chiro= hand, ptera= wing). Bats are more closely related to primates (monkeys and humans) than they are to rodents. There are two families and 18 species of bats in the eastern United States. Fourteen species can be found in Florida. All these bats feed on night-flying insects. Each bat eats about its weight in food every night. This means that even a small colony, numbering several hundred individuals, consumes hundreds of pounds of insects every week. These insectivorous bats have tiny sharp teeth for chewing insects. Bats cannot use their teeth to gnaw wood or wires as can rodents with their chisel-like incisors.

During the day bats rest in dark secluded roosts, such as caves, hollow trees, under bridges, crevices, and the attics of buildings. In winter when insects are scarce, some bats migrate like some birds do, while others hibernate in caves, trees, or buildings. Most bats in Florida enter torpor (a form of deep sleep) during the day and on winter nights when it is too cold for their insect food to fly. Most bat species only have one baby per year. So it takes bat populations a long time to recover from human acts of destruction. Bats are long lived animals. The little brown bat from the northern states is known to live up to 30 years. Bats in Florida can probably live 10 to 12 years. Bats are creatures of habit and will frequent the same roost year after year, even if they only use it seasonally.

Bats are often feared as carriers of rabies. Bats can become infected with the rabies virus as can dogs, cats, raccoons, and skunks. But unlike these animals, rabies-infected bats do not generally become enraged and attack people or other animals. They usually become paralyzed and die quietly. The infection rate for house-dwelling bats is very low, ranging from 1 per 2000 (0.05%) in the southeastern bat to 4 per 1,130 (0.35%) in the Brazilian free-tailed bat.

BAT IDENTIFICATION

A picture key to the bats of Florida is included in this publication. Table 1 summarizes the natural history of Florida bats.

REMOVING SINGLE BATS FROM A BUILDING

Despite their importance as insect predators, bats can be a nuisance when they choose to live in houses, buildings, or other structures used by people. Problems such as noise, smell, accumulations of feces (guano) and urine, staining and spotting of surfaces, attraction of other pests such as flies or cockroaches, and the general fear of these mammals by the public may require that they be excluded from a structure.

Single bats occasionally enter buildings accidentally. This usually occurs in the spring or fall, when bats move between winter roosts and maternity roosts, or in the late summer when young bats have just learned to fly. Young bats can become confused, get lost, and turn up inside buildings where they don't belong. In most cases, all that is required is that access for escape be provided by opening a door or window. In cases where that is not possible, as in most air conditioned buildings, a bat can be captured by covering the resting individual with an empty coffee can. Then gently slide a piece of cardboard or heavy paper between the container and the surface on which the bat is resting, trapping it inside. Small groups of bats, numbering fewer than 10 individuals, can also be removed in this way. If a small bat is resting quietly, it may be possible to pick it up while wearing heavy leather gloves. Never touch a bat with bare hands, it will bite to defend itself, as would any wild animal. Do not try to catch a flying bat; this is almost impossible and usually results in injuring the animal. After the bat is captured, take it outside, away from children and pets and let it fly away or place it high on the side of a tree or wall to fly away

TABLE 1. Bats of Florida

Common Names	Species	Distribution In Florida	Habitat Summer	Habitat Winter	Notes
Southeastern Bat	*Myotis austroriparius*	Northern 2/3 of the state	Caves, Trees, Buildings	Caves, Buildings	Common building inhabitants
Gray Bat	*Myotis grisescens*	Panhandle	Caves	Caves	Endangered Species
Northern Long-eared Bat	*Myotis septemtrionalis*	Panhandle	Hollow Trees, Buildings	Caves	Very rare in Florida
Indiana Bat	*Myotis sodalis*	Panhandle	Hollow Trees, Under loose bark	Caves	Endangered Species. Very rare in Florida
Little Brown Bat	*Myotis lucifugus*	Panhandle	Hollow Trees, Buildings	Caves	Very rare in Florida
Eastern Pipistrel	*Pipistrellus subflavus*	Northern 2/3 of the state	Trees	Caves	Rarely enter houses.
Big Brown Bat	*Eptesicus fuscus*	Northern 1/2 of the state	Buildings and Under Bridges	Caves, attics	Uncommon in Florida
Red Bat	*Lasiurus borealis*	Norhtern 1.3 of the state	Trees (foliage)	Trees (foliage)	Tree bat
Seminole Bat	*Lasiurus seminolus*	All of state except Everglades Region	Tree Foliage, Spanish Moss	Tree Foliage, Spanish Moss	Tree bat

on its own. A torpid (cold and sleepy) bat will need to "warm up" before it can fly.

If anyone is bitten, cleanse the wound thoroughly with soap and water and call your county health department for information and instructions. Try to collect the bat so it can be tested for rabies. NEVER pick up a bat you find lying on the ground. Keep children and pets away from it and if necessary move it to an inaccessible spot with a shovel or similar implement. Call the county animal control office to have the bat removed.

CONFIRMING THE PRESENCE OF BATS IN A BUILDING

The presence of a bat colony in a building is often confirmed by seeing bats emerge from various openings at dusk. Squeaking and rustling noises coming from ceilings and walls may indicate a bat colony is present. The sounds may also come from mice or flying squirrels. Chirping noises coming from chimneys are usually made by nesting chimney swifts, which are small insect-eating birds. An opening, which can be as narrow as 1/4 inch, with a dirty stain below it may be the exit hole for bats. Stains come from urine, feces, and body oils that are deposited around the opening as bats enter or leave the roost. Droppings on sidewalks, ledges, patios, or underneath rafters in an attic or barn may indicate bats are present. Bat droppings, which are brown or black and resemble instant rice grains in size and shape, are composed entirely of insect parts. Mouse droppings are similar in size and shape but do not crumble between your fingers to reveal bits of insects. Gecko droppings are similar to bat droppings but the pieces of insects are larger, less chewed up, and have a small white ball of uric acid on one end. Cockroach droppings are usually smaller and have 6 flattened sides, making them hexagonal in cross section.

BAT PROOFING

As with most nuisance animal situations, preventing a problem is much easier and cheaper than correcting one. To prevent bats from establishing themselves in a building, all attic and soffit vents should be screened with 1/4 inch hardware cloth or screen. Good ventilation of attics discourages bats from roosting and also discourages infestations of large peridomestic cockroaches. Vent holes in spanish tile roofs should be covered with screen that is held in place with silicon rubber chalk. Gaps in siding, spaces under warped fascia boards, spaces between house and chimney, and loose flashing and moldings should be sealed to exclude bats and other invading household pests.

EXCLUDING A BAT COLONY

When bats do become established in a building where they are not wanted, the best and most permanent solution is exclusion. This is accomplished by the following steps:

1. Observe the building at dusk from all angles on three or four successive evenings to identify the entrance and exit openings that the bats are using.

2. Seal and bat-proof all other openings that bats do not use, but might use in the future. Some species of bats can enter through a crack or crevice that is only 1/4 inch wide. Sealing materials can include caulking, wood, sheet metal, plaster, cement, 1/4 inch hardware cloth, or window screen.

3. Plan to do the exclusions in the spring or fall. Bats give birth in the summer. Exclusions must not be attempted when baby bats are present, as they do not fly with their mothers until they are almost full grown. Baby bats trapped in the roost by an exclusion will die of thirst or starvation and create a serious odor and fly problem. In Florida, exclusions should not be attempted from May through August. Wait until the young are flying to exclude the colony. Avoid exclusions during cold weather because bats usually do not fly when temperatures are below 45° F.

4. Exclude the unwanted bats by placing one-way devices on the colony's exit points. These devices can be as simple as a plastic "sleeve". Once the bats exit through this, it collapses behind them. They can not climb or crawl on the smooth plastic. Bat netting works the same way. The top of the netting is attached securely to a wall, beam, or other solid surface above the roost opening and extends over it. The bottom of the netting is secured at spots along the bottom edge. The netting can be secured with duct tape, staples, velcro tabs, or silicon rubber chalking. The bats exit the roost, crawl out the bottom of the netting, escape, but are not able to find the roost opening when they return from feeding, because the netting covers the hole. Professional bat exclusion specialists have developed a variety of exclusion devices for special situations. Colonies in large structures or in high dangerous places should be excluded by experienced professionals. Returning bats may fly around the roost openings, but will disperse within a day or so.

5. Once excluded, a large bat colony may leave behind external parasites such as bat bugs, soft ticks, or mites. Most bat parasites are host specific and will not bite people. Once the bats have been excluded, the application of a desiccant

or insecticide dust throughout the roosting site will kill parasites. This is a good precaution to prevent their spread while they look for other hosts.

Bat guano dries to form a crumbly, powdery substance that can grow a fungus called *Histoplasma capsulatum.* Spores from this fungus become airborne when the guano is disturbed. Inhaled spores develop into a yeast-like infection in the lungs. This produces a systemic disease called histoplasmosis, the effects of which can range from flu-like symptoms (in most people) to serious lung abscesses and lesions resembling tuberculosis (in a minority of others). When working in an area where bat guano is present wear protective clothing and a cartridge respirator (capable of filtering particals as small as 2 microns) to avoid breathing guano dust. Prior to removing accumulated guano, spray it with a 1:10 bleach and water solution to hold down the dust and kill the fungus.

6. Permanently seal roost openings when you are sure all bats have left the roost. Leave the excluder in place for at least three days in warm weather, longer in cool or cold weather.

OTHER METHODS OF BAT EXCLUSION

Bright lights can be used to discourage bats from roosting in large structures that are difficult to seal, such as warehouses, barns, or similar buildings. Fiberglass insulation also discourages bats from roosting; this is probably due to the irritating nature of this material. Ultrasonic sound emitters for control of bats are expensive ($20 to $70) and there is no scientific evidence to indicate that they actually work. The animals simply move into sound shadows to avoid the sound.

BATS AND THE LAW

Two species in Florida (the Indiana bat and the gray bat, which generally do not occur in buildings) are classified as endangered species by the U. S. Fish and Wildlife Service. In Florida, these and all other bats are classified as native nongame wildlife by the Florida Game and Fresh Water Fish Commission and are protected by law. Title 39-12.009 of the Florida Wildlife Code (effective date July 1, 1992) states that nuisance or destructive wildlife may not be taken with a gun and light, steel traps, or poisons without a specific permit issued by the executive director of the FG&FWFC. Since NO POISONS OR FUMIGANTS ARE LEGAL OR REGISTERED FOR CONTROL OF BATS IN FLORIDA, the FG&FWFC cannot issue any permits to poison bats. Poisoning a bat colony exposes people, especially children, and pets to large numbers of dead and dying bats, thus increasing the chance of someone being bitten by picking up a sick animal. Exclusion is the only recommended permanent solution to an unwanted bat colony in a building and does not require a nuisance animal permit. A nuisance wildlife permit is required to capture or kill any bats.

Naphthalene repellents are the only registered materials for bat control in Florida. However, use of this substance is not a permanent solution. Naphthalene evaporates and as soon as this occurs, the bats will return unless roost openings have been sealed. There is also the odor and expense of placing several pounds of naphthalene in a building where people live. If people are sensitive to the odor of moth balls, avoid using naphthalene. If naphthalene is used to repel bats from a structure, a nuisance animal permit is recommended to protect the user legally in the event that bats are accidently killed.

BAT HOUSES

Since bats are beneficial, many people want to keep them in the neighborhood while excluding them from a building. This may be accomplished by putting up one or more commercially available bat houses. Bat houses are similar to bird houses in size and shape. However, they lack a circular entry hole. Bats enter and exit the house from the bottom of the house, which has been left open. Bats will not leave a building to move into a bat house. But if a colony is excluded, bats may move into the closest new accommodations. A bat house should be placed high on the east side of a tree or building (10 to 15 ft.), as far in advance of the exclusion as possible. There should be no branches directly below the bat house to obstruct the entry or exit of bats. Plans for a bat house are available from Bat Conservation International, P.O. Box 162603, Austin, Texas, 78716-2603.

Birds

English or house sparrows (*Passer domesticus*), starlings (*Sturnus vulgaris*), rock dove or domestic pigion (*Columba livia*), and the muscovy duck (*Cairina moschata*) are unprotected exotic birds. They may be shot, live trapped, snared or captured by hand without any permit or license. Local ordinances may prohibit discharge of firearms and may protect all birds. The use of the avicide, Avitrol™, for pigeon control, requires a Florida Avitrol Permit available from the Division of Wildlife, Florida Game & Fresh Water Fish Commission (F.G. & F.W.F.C.). Care must be taken to prevent accidently poisoning nontarget native birds. A permit is not required to use porcupine wire, tactile roost repellents, exclusion netting, and startle devices such as propane cannons, artificial snakes, models of hawks or owls, etc.

Blackbirds (*Agelaius spp.* and *Euphagus spp.*), grackles (*Quiscalus spp.*), cowbirds (*Molothrus spp.*), and crows (*Corvus spp.*) may be taken by a property owner without a permit when they are in the act of or about to damage ornamental trees, agricultural crops, wildlife, livestock, or are concentrated so as to be a threat to human health. The use of the avicide Avitrol™, requires a Florida Avitrol Permit. These birds are protected when not causing damage. When crows are not causing damage, they may only be taken during the legal crow hunting seasons. A hunting license is required to hunt crows. If pest control operators or others are hired by the property owner to remove these species of nuisance birds, they must first acquire a Nuisance Animal Permit from Regional Offices of the F.G. & F. W. F. C.

All other species of birds are protected under Federal Law and International Treaty. When these migratory birds become a nuisance, permits from the F.G. & F.W.F.C. and United States Fish and Wildlife Service is required prior to taking or attempting to take these birds or their nests. Birds that fall in this category include the following examples; chimney swifts nesting inside chimneys, Carolina wrens nesting inside buildings, woodpeckers nesting and pecking, vultures roosting, flocks of gulls around garbage dumps, airports, warehouses, and parking lots. For advice or recommendations on dealing with nuisance migratory birds contact USDA Animal Damage Control Program Office in Tallahassee, Florida.

Common Florida Spiders

J. L. Castner[1]

Spiders belong to the class Arachnida, which contains organisms with four pairs of legs, no antennae, and two body regions. A shield-like carapace covers the head and area from which the legs arise. Their mouthparts function vertically and are called chelicerae.

Jumping spiders. The jumping spiders belong to the family Salticidae and are sometimes called salticids. All species are small, usually less than 15mm long. They are easily identified by their eye arrangement, which is in three rows. Jumping spiders do not construct webs, but actively hunt prey during the day, pouncing on their luckless victims. Many are brightly colored, sometimes with iridescent chelicerae (mouthparts) as in the genus *Phidippus*. Some species are commonly found on or around buildings, such as *Plexippus* spp.

Crab spiders. Crab spiders are so named because they hold their legs to the side in a crab-like fashion. They commonly range from 5 to 10mm in length. These spiders do not spin webs, but wait in ambush on flowers and foliage for their insect prey. Crab spiders such as *Misumenoides* spp. are often extremely well camouflaged, blending in perfectly with the flowers among which they live.

Golden silk spider, *Nephila clavipes*. The golden silk spider is found throughout Florida and the southeastern U.S. The female is distinctively colored and among the largest orb-weaving spiders in the country. The female varies from 25 to 40mm long and has conspicuous hair tufts on her long legs. Males are small, approximately 4 to 6mm in length, dark brown in color, and often found in the webs of females. These spiders feed primarily on flying insects which they catch in webs that may be greater than a meter in diameter. They are most commonly found in forests, along trails and at clearing edges.

Spiny orb-weaver, *Gasteracantha cancriformis*. The spiny orb-weaver spider is one of the most colorful and easily recognized spiders in Florida. The dorsum of the abdomen is usually white with black spots and large red spines on the margin. Females range from 5 to 10mm long and 10 to 14mm wide. The webs typically contain tufts of silk, which may prevent birds from flying into them.

Black and yellow argiope spider, *Argiope aurantia*. The argiope spiders are a large and distinctive group. Their large, conspicuous webs can often be seen along the edge of woodlands. The black and yellow argiope can reach a length of 25mm. Its characteristic silver carapace and yellow and black markings make it easy to identify. Argiope spiders tend to hang head down in the middle of a medium-sized web that has thickened, zigzag bands of silk in the center.

Green lynx spider, *Peucetia viridans*. This spider is commonly encountered on shrubs, weeds, and foliage. The female ranges from 12 to 20mm in length, while the male seldom gets beyond 12mm. The body is a vivid, almost transparent green, with red spots and some white markings. The legs are long, slender and covered at intervals with long, black spines. These spiders have good eyesight and hunt and stalk their prey during the daytime. They spin no webs, but sometimes anchor themselves with silk. They are important predators of caterpillar pests of row crops.

Wolf spiders. Wolf spiders belong to the family Lycosidae. They are very common and usually found on the ground, where they are well camouflaged. The Carolina wolf spider (*Lycosa carolinensis*) is the largest in the U.S. with a size of 25 to 35mm. These spiders do not spin webs, but some dig burrows or hide under debris. Like other hunting spiders, they have good eyesight and are sensitive to vibrations.

Long-jawed orb-weavers, genus *Tetragnatha*. These spiders have the characteristic behavior of clinging to a support with their short third pair of legs while holding their remaining and much longer legs in an extended manner both in front of and behind the body. They spin small webs that are 8 to 12" in diameter and catch small flying insects. They are often found in association with foliage bordering water.

---

1. Scientific Photographer, Department of Entomology and Nematology, Cooperative Extension Service, Institute of Food and Agricultural Sciences (IFAS), University of Florida, Gainesville.

COOPERATIVE EXTENSION SERVICE, UNIVERSITY OF FLORIDA, INSTITUTE OF FOOD AND AGRICULTURAL SCIENCES, John T. Woeste, Director, in cooperation with United States Department of Agriculture, publishes this information to further the purpose of the May 8 and June 30, 1914 Acts of Congress; and is authorized to provide research, educational information and other services only to individuals and institutions that function without regard to race, color, sex, age, handicap or national origin. Information on copies for purchase is available from C. M. Hinton, Publications Distribution Center, IFAS Building 664, University of Florida, Gainesville, Florida 32611. Printed 11/92.

Common Florida Spiders

Jumping spider, *Phidippus*

Jumping spider, *Plexippus*

Crab spider

Crab spider

Golden silk spider

Spiny orb-weaver

Black and yellow argiope spider

Green lynx spider

Carolina wolf spider

Long-jawed orb-weaver

Index